The Simple Science of Flight

DATE DUE

DEMCO 38-296

The Simple Science of Flight

From Insects to Jumbo Jets

revised and expanded edition

Henk Tennekes

The MIT Press
Cambridge, Massachusetts
London, England

For information about special quantity discounts, email specialsales@mitpress.mit.edu.

Set in Melior and Helvetica Condensed on 3B2 by Asco Typesetters, Hong Kong. Printed and bound in the United States of America.

Library of Congress Cataloging-in-Publication Data

Tennekes, H. (Hendrik)
The simple science of flight : from insects to jumbo jets / Henk Tennekes. — Rev. and expanded ed.
    p.   cm.
Includes bibliographical references and index.
ISBN 978-0-262-51313-5 (pbk. : alk. paper)
1. Aerodynamics. 2. Flight. I. Title.
TL570.T4613 2009
629.132'3—dc22   2009012431

10 9 8 7 6 5 4 3 2 1

to my grandchildren, Nick and Emma

Contents

White-fronted goose (*Anser albifrons*): $W = 17$ N, $S = 0.18$ m^2, $b = 1.40$ m.

Preface

I wrote a preliminary draft for this book in 1990. Much has happened since. For one, the book generated a lot of response, both from professionals in a variety of disciplines and from interested outsiders. Professional criticism concentrated on my decision not to delve into the details of aerodynamics. I am a turbulence specialist; I could easily fill an entire book with the curious tricks airflows can play. But that would lead me astray and confuse my readers. In the first edition I stuck to flight performance; in this revision I maintain that choice, except for a novel treatment of the aerodynamics of induced drag and trailing vortices in chapter 4.

I much enjoyed using parts of my book in college-level courses for senior citizens. The always lively interaction with my audience forced me to ponder how I should avoid pitfalls, where I should explain things in more detail, where I should tighten the argumentation, and what I should leave out. Similar feedback on the lecture circuit also gave me plenty food for thought. One frequent source of misunderstanding was my rather casual use of numbers. In this book I am not interested in great accuracy. I much prefer representative approximate numbers above three-digit precision. I want my book to be accessible to a wide audience; the nitpicking typical of much work in the so-called exact sciences would make it harder to achieve this objective.

I remember vividly how I was corrected in one of my senior citizens' courses. I was talking about bicycle racers and pilots of human-powered aircraft, and I explained that their hearts grow bigger from continuous exercise, just like their leg muscles. At one point, a gentleman in the last row raised his hand and said "Henk, it isn't just heart size that counts; the entire circulatory system is adjusted to long-distance performance." He turned out to be a retired professor of cardiology, and he proceeded with a twenty-minute lecture on hearts, the elasticity of arteries, and what have you. The class loved it, and so did I. But the incident was also a reminder that I shouldn't get stuck in matters outside my field of competence. *The Simple Science of Flight* is meant for a general audience; it is neither a vehicle for scientific debates on animal physiology nor a manual for airplane design.

Still, I admit I am fascinated by the similarities between nature and technology. I learn by association, not by dissociation. Swans and airliners follow the same aerodynamic principles. Biological evolution and its technological counterpart differ in many ways, but I find the parallels between them far more exciting. Just what is it that makes some airplanes successful, but others misfits? And can such matters be explained without sophisticated advanced mathematics? Would high school algebra suffice?

Many popular science books make quite a fuss of the marvelous progress of science and engineering. All too often the hidden message is "Dear reader, you are but a layperson. You should have deep respect for the sophistication revealed to you by specialists, who are the ones who really understand the secrets of the universe, the building blocks of life, the fantastic blessings of computer technology, or the great achievements of aerospace engineering." I have never been a fan of such grandiose perspectives. They tend to elevate science to a level where ordinary people have no choice but to kneel. I prefer to make science accessible, and I don't mind that it takes a lot of effort to struggle with the arcane texts presented in scientific journals. Also, I do not agree that one's respect for miracles is lessened by an attempt to understand them. On the contrary, one's sense of wonder can only grow as one's insight increases. After one has computed how large a swallow's wings should be, one's respect for the magnitude of the mystery that keeps the bird in the air can only be greater. The intimate knowledge of meteorology that migrating monarch butterflies apparently possess helps me to keep my feet on the ground.

There is a lot of news from the research front. Systematic observation by several groups of ornithologists during the last twenty years has made it clear that the migration speeds of many birds are substantially higher than the speed calculated from the simple algorithms I had condensed from the professional literature. I used to think that only homing pigeons fly at top speed regardless of the consequences. (They race at more than 40 miles per hour, though taking it easy at 25 mph would minimize their fuel consumption.) Now it appears that all migratory birds fly as fast as their muscles allow when they are in a hurry. Not all species are in a hurry, though; the migratory habits of gulls and terns, for example, seem rather relaxed.

The study of bird migration has made giant strides since 1990, thanks primarily to the continuing accumulation of radar data but also thanks to the use of lightweight transmitters and fieldwork on the Arctic tundra. The current migration champion is the bar-tailed godwit (*Limosa lapponica*), a fairly large wading bird that weighs 500 grams at takeoff. It has been confirmed by several parties that the godwit flies nonstop across the Pacific Ocean, from Alaska to New Zealand—a distance of 11,000 kilometers (7,000 miles). That feat, comparable to the performance of a long-distance airliner, proves that the godwit has much better aerodynamics and much better muscle efficiency than was previously thought, and that it undergoes rather severe physiological adaptations before and during its seven-day flight. New evidence on the migration of other wader species points in the same direction. Professionals have severely underestimated birds' flight performance. Impressed by the new evidence, I had to make many changes throughout the book. New wind-tunnel studies also have generated excitement. The champion of those studies is a young female jackdaw (*Corvus monedula*) that apparently was quite at ease in the wind tunnel of the Flyttnings Ekologi (Flight Ecology) department of the Ekologiska Institutionen at Lund University in Sweden. It exhibited superb gliding performance notwithstanding its rather ordinary wings. It would glide more than 12 feet forward for each foot of height loss, twice as far as most researchers had thought. In retrospect, many of the early wind-tunnel experiments with birds and even some recent ones failed to produce reliable results.

In the first edition I dealt with flapping flight in a rather offhand way, partly because there were very few wind-tunnel data available. This time I can be a lot more specific. The Swedish jackdaw receives considerable attention, because its best gliding speed is much lower than its reported migration speed.

I was a fan of the Boeing 747 when I conceived the first edition of this book, and I remain a dedicated fan of the Big Bird that, in the 1969 phrasing of *Newsweek*, introduced A New Air Age. But much has happened since. Airbus, the European conglomerate, now markets the A380, which is meant to drive the 747 into oblivion. And Boeing has responded. Its 777 offers transportation capacity equivalent to that of the 747, but with improved aerodynamics and superior engine efficiency. In the first edition, I wrote that the Boeing 747 had been the dominant mode of intercontinental transportation for 25 years, and that it would remain in service for at least another 25. Yes, it will. However, like an old warrior, it will fade away 30 years from now. Other airplanes, such as the Boeing 777 and 787 and the Airbus A350, will promote the idea that we don't have to change planes as we fly from Hamburg to Pittsburgh. The hub-and-spoke system of airports will no longer dominate intercontinental traffic.

The Concorde went out with a bang. A fiery crash near Paris on July 25, 2000, signaled the end of its career. It didn't quite make the centennial of the Wright brothers' first powered flight. In chapter 6, I reflect on the fate of supersonic transportation. In retrospect, the Concorde was a fluke, more so than anyone could have anticipated. From an evolutionary perspective it was a mutant. It was a very elegant mutant, but it was only marginally functional. The fate of the Concorde inspired me to draw parallels between biological evolution and its technological counterpart wherever appropriate.

The Simple Science of Flight has been my sweetheart ever since I started dreaming of it, back in 1978. It has become a favorite of many readers all over the world. Its revitalization and rejuvenation will surely endear it to the next generation of people who are, like me, enthralled by everything that flies.

The Simple Science of Flight

Laughing gull (*Larus atricilla*): $W = 3.3$ N, $S = 0.11$ m^2, $b = 1.03$ m.

1 Wings According to Size

Imagine that you are sitting in a jumbo jet, en route to some exotic destination. Half dozing, you happen to glance at the great wings that are carrying you through the stratosphere at a speed close to that of sound. The sight leads your mind to take wing, and you start sorting through the many forms of flight you have encountered: coots and swans on their long takeoff runs, seagulls floating alongside a ferry, kestrels hovering along a highway, gnats dancing in a forest at sunset. You find yourself wondering how much power a mallard needs for vertical takeoff, and how much fuel a hummingbird consumes. You remember the kites of your youth, and the paper airplane someone fashioned to disrupt a boring class. You recall seeing hang gliders and parawings over bare ski slopes, and ultralights on rural airstrips.

What about the wings on a Boeing 747? They have a surface area of 5,500 square feet, and they can lift 800,000 pounds into the air—a "carrying capacity" of 145 pounds per square foot. Is that a lot? A 5 × 7-foot waterbed weighs 2,000 pounds, and the 35 square feet of floor below it must carry 57 pounds per square foot—almost half the loading on the jet's wings. When you stand waiting for a bus, your 150 pounds are supported by shoes that press about 30 square inches (0.2 square foot) against the sidewalk. That amounts to 750 pounds per square foot—5 times the loading on the jet's wings. A woman in high heels achieves 140 pounds per square inch, which is 20,000 pounds per square foot.

From a magazine article you read on a past flight, you recall that a Boeing 747 burns 12,000 liters of kerosene per hour. A hummingbird consumes roughly its own weight in honey each day—about 4 percent of its body weight per hour. How does that compare to the

747? Midway on a long intercontinental flight, the 747 weighs approximately 300 tons (300,000 kilograms, 660,000 pounds). The 12,000 liters of kerosene it burns each hour weigh about 10,000 kilograms (22,000 pounds), because the specific gravity of kerosene is about 0.8 kilogram per liter. This means that a 747 consumes roughly 3 percent of its weight each hour.

A hummingbird, however, is not designed to transport people. Perhaps a better comparison, then, is between the 747 and an automobile. At a speed of 560 miles per hour, the 747 uses 12,000 liters (3,200 U.S. gallons) of fuel per hour—5.7 gallons per mile, or 0.18 mile per gallon. Your car may seem to do a lot better (perhaps 30 miles per gallon, or 0.033 gallon per mile), but the comparison is not fair. The 747 can seat up to 400 people, whereas your car has room for only four. What you should be comparing is fuel consumption per *passenger*-mile. A 747 with 350 people on board consumes 0.016 gallon per passenger-mile, no more than a car with two people in it. With all 400 seats occupied, a 747 consumes 0.014 gallon per passenger-mile. A fully loaded subcompact car consuming 0.025 gallon per mile (40 miles per gallon) manages 0.006 gallon per passenger-mile.

Nine times as fast as an automobile, at comparable fuel costs: no other vehicle can top that kind of performance. But birds perform comparable feats. The British house martin migrates to South Africa each autumn, the American chimney swift winters in Peru, and the Arctic tern flies from pole to pole twice a year. Birds can afford to cover these enormous distances because flying is a relatively economical way to travel far.

Lift, Weight, and Speed

A good way to start when attempting to understand the basics of flight performance is to think of the weight a pair of wings can support. This "carrying capacity" depends on wing size, airspeed, air density, and the angle of the wings with respect to the direction of flight.

Unfortunately, most of us learned in high school that one needs the Bernoulli principle to explain the generation of lift. Your science teacher told you that the upper face of a wing has to have a convex curvature, so that the air over the top has to make a longer journey than that along the bottom of the wing. The airspeed over the top of the wing has to be faster than that below, because the air over the top "has to rejoin" the air along the bottom. An appeal to Bernoulli then "proves" that the air pressure on top is lower than that below. The biologist Steven Vogel, who has written several delightful books on biomechanics, says: "A century after we figured out how wings work, these polite fictions and misapprehensions still persist." Polite fiction, indeed. It does not explain how stunt planes can fly upside down, it does not explain how the sheet-metal blades of a home ventilator or an agricultural windmill work, it does not explain the lift on the fabric wings of the Wright Flyer, it fails to explain the aerodynamics of paper airplanes and butterfly wings, and so on. If your high school teacher had taken the trouble to do the math, he would have found that the mistaken appeal to Bernoulli does not produce nearly enough lift to keep a bird or an airplane aloft. The principal misapprehension in the conventional explanation is that the air flowing over the top of a wing has to rejoin the air flowing along the bottom when it reaches the trailing edge. In fact, all along the wing the airspeed over the top is higher than that over the bottom. Rejoining is not necessary and does not occur.

We will have to do better. I will use a version of Newton's Second Law of Motion, not familiar to most high school physics teachers, that is a cornerstone of aerodynamics and hydrodynamics. I also will appeal to Newton's Third Law, which says that action and reaction are equal and opposite. Applied to wings, these two laws imply that a wing produces an amount of lift that is equal to the downward impulse given to the surrounding air. According to the version of the Second Law that I will use, force equals rate of change of momentum and can be computed as mass flow times speed change.

How much air flows around a wing? The mass flow is proportional to the air density ρ, the wing area S, and the airspeed V. Let's check the dimensions of the product of the three factors ρ, V, and S. The density ρ is measured in kilograms per cubic meter, the wing area S (taken as the planform surface seen from above) in square meters, and the speed V in meters per second. This means that the units for ρVS are kilograms per second, which indeed is a mass flow. For a Boeing 747-400 cruising at 39,000 feet, the mass flow around the wings computes as 42 tons of air per second, or 2,500 tons per minute. By the way, the mass flow into each of a 747-400's jet engines is about 500 pounds per second.

How much downward motion is imparted to the air flowing around a wing? The downward component of the airspeed leaving the wing is proportional to the flight speed (V) and the angle of attack of the wing (α). It is easy to get a feeling for the effect of the angle of attack: just stick your hand out of the window of a car moving at speed. When you keep your hand level you feel only air resistance, but when you turn your wrist your hand wants to move up or down. You are now generating aerodynamic lift. Note also that you start generating more resistance while losing much of the lift when you increase the angle of your hand in the airstream. Airplanes and birds have similar problems: when the angle of attack of their wings reaches about 15°, the air flow over the top surface is disrupted. Pilots call this "stall." When the airflow is stalled, the lift decreases; it is no longer proportional to the angle of attack. On top of that, the drag increases a lot, causing a plane to drop like a brick.

With the mass flow pinned down as ρVS and with the deflection speed proportional to the product of α and V, the lift on a wing is proportional to $\alpha\rho V^2 S$. Note that the *square* of the airspeed V is involved. When you fly twice as fast with the same wings at the same angle in the air flow, you obtain 4 times as much lift. You'll have to reduce the angle of attack if you merely need to support your weight, or you may decide to make a tight turn. At an altitude of 12 kilometers, where the air density is only one-fourth its sea-level value, you will have to fly twice as fast to sustain your weight.

What about Bernoulli? The conventional explanation is that the air over the top surface has to flow faster than the air below, so that the pressure on the top surface will be lower than that along the bottom surface. That "logic" is inverted. A wing gives the surrounding air a downward deflection. It does so by creating a region of reduced pressure on the top surface (a kind of "suction"), which pulls the passing air downward. The partial vacuum over the top surface manifests itself as lift. Yes, the suction over the top accelerates the local airflow, and yes, the pressure difference can be computed with the Bernoulli formula, but the "polite fictions" involved in what you learned in high school lead you astray.

Birds and airplanes can change the angle of attack of their wings to fit the circumstances. They fly nose up, with a high angle of attack, when they have to fly slowly or have to make a sharp turn; they fly nose down when speeding or diving. But everything that flies uses about the same angle of attack in long-distance cruising; $6°$ is a reasonable average. At higher angles of attack the aerodynamic drag on wings increases rapidly; at smaller angles wings are underutilized.

Since wings have to support the weight of an airplane or a bird against the force of gravity, the lift L must equal the weight W. The lift is proportional to the wing area S and to ρV^2, and so is the weight:

$$W = 0.3\rho V^2 S. \tag{1}$$

(The 0.3 is related to the angle of attack in long-distance flight, for which the average value of $6°$ has been adopted.)

We must make sure we aren't violating the rules of physics when we use equation 1. We must give clear and mutually consistent definitions for the units in which ρ, V, and S are expressed. (Clearly the numbers would look different if velocities were given in miles rather than in millimeters per minute.) The best way to ensure consistency is to use the metric system, expressing S in square

Great tit (*Parus major*): $W = 0.2$ N, $S = 0.01$ m^2, $b = 0.23$ m.

meters, V in meters per second, and ρ in kilograms per cubic meter. The rules of physics then require that the weight W in equation 1 be given in kilogram-meters per second squared. This frequently used unit is known as the *newton*, after Sir Isaac Newton (1642–1727), the founder of classical mechanics. A newton is slightly more than 100 grams (3.6 ounces). A North American robin weighs about 1 newton, a common tern a little bit more, a starling a little bit less. Since there are roughly 10 newtons to a kilogram, a 70-kilogram (154-pound) person weighs about 700 newtons.

The distinction between mass and weight still causes confusion in the public mind. Mass is the amount of matter; weight is the downward force that all matter experiences in Earth's gravity field. One reason for the confusion is that the force of gravity is proportional to the mass of an object and is independent of everything else. None of the other forces in nature have this outrageously simple property. I have chosen to work with the weight of flying objects, not their mass, because all flying has to be done on Earth and is therefore subject to terrestrial gravity. If gravity were absent, wings would not be needed. Classical Italian painters understood this well: their Cupids, being little angels, feature miniature wings, mere adornments. Angels need not worry about gravity.

Sparrow hawk (*Accipiter nisus*): $W = 2.5$ N, $S = 0.08$ m^2, $b = 0.75$ m.

If we respect the rules, we can play with equation 1 in whatever way we want. For example, a Boeing 747-200 has a wing area of 5,500 square feet (511 square meters) and flies at a speed of 560 miles per hour (900 kilometers per hour; 250 meters per second) at an altitude of 12 kilometers (40,000 feet), where the air density is only one-fourth its sea-level value of 1.25 kilogram per cubic meter. Using $\rho = 0.3125$ kilogram per cubic meter, $V = 250$ meters per second, and $S = 511$ square meters, we calculate from equation 1 that W must equal 2,990,000 newtons. Because a newton is about 100 grams, this corresponds to approximately 300,000 kilograms, or 300 tons. That is indeed the weight of a 747 at the midpoint of an intercontinental flight. At takeoff it is considerably heavier (the maximum takeoff weight of a 747-200 is 352 tons), but it burns 10 tons of kerosene per hour.

Equation 1 can be used in several ways. Consider a house sparrow. It weighs about an ounce (0.3 newton), flies close to the ground (so that we can use the sea-level value of ρ, 1.25 kilogram per cubic meter), and has a cruising speed of 10 meters per second (22 miles per hour). We can use equation 1 to find that the sparrow needs a wing area of 0.01 square meter, or 100 square centimeters. That's 20 centimeters from wingtip to wingtip, with an average

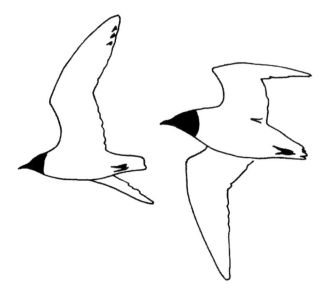

Little gull (*Larus minutus*).

width of 5 centimeters. Or we can use the same equation in designing a hang glider. Taken together, the pilot and the wing weigh about 1,000 newtons (100 kilograms, 220 pounds). So if you want to fly as fast as a sparrow (20 miles per hour), you need wings with a surface area of 33 square meters. On the other hand, if you want to fly at half the speed of a sparrow, your wing area must be more than 100 square meters (more than 1,000 square feet).

Wing Loading

To make equation 1 easier to work with, let us replace the variable ρ (air density) with its sea-level value: 1.25 kilogram per cubic meter. This should not make any difference to most birds, which fly fairly close to the ground. For airplanes and birds flying at higher altitudes, we will have to correct for the density difference or return to equation 1; we can worry about that detail when it becomes necessary. Another improvement in equation 1 is to divide both sides by the wing area S. The net result of these two changes is

Razorbill (*Alca torda*): $W = 8$ N, $S = 0.038$ m^2, $b = 0.68$ m.

$$W/S = 0.38V^2. \tag{2}$$

This formula tells us that the greater a bird's "wing loading" W/S, the faster the bird must fly. Within the approximations we are using here, sea-level cruising speed depends on wing loading only. No other quantity is involved. This is the principal advantage of equation 2. But it is a simplification.

The predecessor of the Fokker 50 was the Fokker Friendship, with a weight of 19 tons (190,000 newtons) and a wing area of 70 square meters. Its wing loading was 2,700 newtons per square meter, good for a sea-level cruising speed of 85 meters per second (190 miles per hour). The wing loading of a Boeing 747 is about 7,000 newtons per square meter, and it must fly a lot faster to remain airborne. The wing loading of a sparrow is only 38 newtons per square meter, corresponding to a cruising speed of 10 meters per second (22 miles per hour). From these numbers one gets the impression that wing loading might be related to size. If larger birds have higher wing loadings, it is no coincidence that a Boeing 747 flies much faster than a sparrow.

Our understanding of the laws of nature is due in part to people who have been driven by the urge to investigate such questions. One person in particular deserves to be mentioned: Crawford H. Greenewalt, a chemical engineer who was chairman of the board of DuPont and a longtime associate of the Smithsonian Institution. For many years Greenewalt's chief hobby was collecting data on the weights and wing areas of birds and flying insects. Humming-birds were his favorites, and he carried out many strobe-light experiments to measure their wing-beat frequencies.

Some of the data collected by Greenewalt and later investigators are listed in table 1. For the sake of clarity, the selection is restricted to seabirds: terns, gulls, skuas, and albatrosses. Looking at table 1, we find that wing loading and cruising speed generally increase as birds become heavier. But the rate at which this hap-pens is not spectacular. A wandering albatross is 74 times as heavy as a common tern, but its wing loading is only 6 times that of its small cousin, and it flies only 2.5 times as fast (equation 2). In terms of weight, the wing loading isn't terribly progressive.

To improve our perception of what is happening, let us plot the weights and wing loadings of table 1 in a proportional or "double-logarithmic" diagram, which preserves the relative proportions be-tween numbers. In a proportional diagram a particular ratio (a two-fold increase, say) is always represented as the same distance, no matter where the data points are located. Four is 2×2, and 100 is 2×50; in a proportional diagram the distance between 2 and 4 is equal to the distance between 50 and 100. (See figure 1.)

The steeply ascending line in figure 1 suggests that there must be a simple relation between size and wing loading. There are devia-tions from this line, of course; for example, the fulmar has a rather high wing loading for its weight. But before you look at the excep-tions, let me explain the rule.

Table 1 Weight, wing area, wing loading, and airspeeds for various seabirds, with W given in newtons (10 newtons = 1 kilogram, roughly), S in square meters, and V in meters per second and miles per hour. The values of W and S are based on measurements; those for V were calculated from equation 2. In general, larger birds have to fly faster.

	W	S	W/S	V m/sec	mph
Common tern	1.15	0.050	23	7.8	18
Dove prion	1.70	0.046	37	9.9	22
Black-headed gull	2.30	0.075	31	9.0	20
Black skimmer	3.00	0.089	34	9.4	21
Common gull	3.67	0.115	32	9.2	21
Kittiwake	3.90	0.101	39	10.1	23
Royal tern	4.70	0.108	44	10.7	24
Fulmar	8.20	0.124	66	13.2	30
Herring gull	9.40	0.181	52	11.7	26
Great skua	13.5	0.214	63	12.9	29
Great black-billed gull	19.2	0.272	71	13.6	31
Sooty albatross	28.0	0.340	82	14.7	33
Black-browed albatross	38.0	0.360	106	16.7	38
Wandering albatross	87.0	0.620	140	19.2	43

All gulls and their relatives look more or less alike, with long, slender wings, pointed wingtips, and a beautifully streamlined body with a short neck and tail; however, they vary considerably in size. Now compare two types of gull, one having twice the wingspan of the other. If the larger of the two is a scaled-up version of its smaller cousin, its wings are not only twice as long but also twice as wide, making its wing area 4 times as large. The same holds for weight. Because weight goes as length times width times

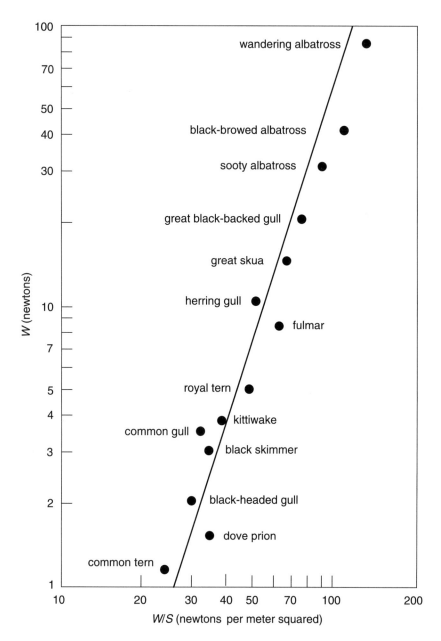

Figure 1 The relation between weight and wing loading represented in a proportional diagram. When the weight increases by a factor of 100, the value of W/S increases by a factor of 5 and the airspeed by a factor of more than 2.

Herring gull (*Larus argentatus*): $W = 11.4$ N, $S = 0.2$ m², $b = 1.34$ m.

height, the weight of the larger gull is 8 times that of its smaller cousin. Eight times as heavy, with a wing area 4 times as large, a bird with a wingspan twice that of its smaller cousin has twice the wing loading. And according to equation 2 it has to fly 40 percent faster (the square root of 2 is about 1.4). It is useful to write this down in an equation. If the wingspan (the distance from wingtip to wingtip with wings fully outstretched) is called b, the wing area is proportional to b^2 and the weight is proportional to b^3. The wing loading, W/S, then is proportional to b. But b itself is proportional to the cube root of W. In this way we obtain the scale relationship

$$W/S = c \times W^{1/3}. \tag{3}$$

Strictly speaking, this formula holds only for birds that are "scale models" of one another. The steeply ascending line in figure 1 corresponds to equation 3, the coefficient having been determined empirically. For the seabirds in figure 1, $c = 25$: at a weight of 1 newton, the wing loading is 25 newtons per square meter.

The scale relation (equation 3) is universally applicable whenever weights and supporting surfaces or cross-sectional areas are involved. Galileo Galilei (1564–1642) wrote the first scientific treatise on this subject, asking himself why elephants have such thick legs and similar questions. The answer is that the larger an animal gets, the more crucial the strength of the legs becomes. The stress on leg bones increases as the cube root of weight; for this reason, a land animal much larger than an elephant is not a feasible proposition. This is the same problem that engineers face when they design bridges, skyscrapers, or even stage curtains, which would give way under their own weight were they not reinforced by steel cables. Another good example is that of walking barefoot on a stony beach. Walking on gravel is an uncomfortable experience for adults, but not for little children. A father who is twice as tall and 8 times as heavy as his 8-year-old daughter must support himself on feet whose surface area is only 4 times that of her feet. Thus, his "foot loading" is twice hers. No wonder he seems to be walking on hot coals.

The scale relation given in equation 3 is not a hard-and-fast rule. Most birds are not exact "scale models" of others, and we must also allow some latitude for deviations to fit designers' visions and nature's idiosyncrasies. On the other hand, designers are confronted by tough technical problems whenever they deviate too far. The margins permitted by the laws of scaling are finite.

The Great Flight Diagram

Thanks to the dedicated work of Crawford Greenewalt and other enthusiasts, and assisted by the airplane encyclopedia *Jane's All the World's Aircraft*, we can now collect everything that flies in a single proportional diagram: figure 2. The results are impressive: 12 times a tenfold increase in weight, 4 times a tenfold increase in wing loading, and 2 times a tenfold increase in cruising speed!

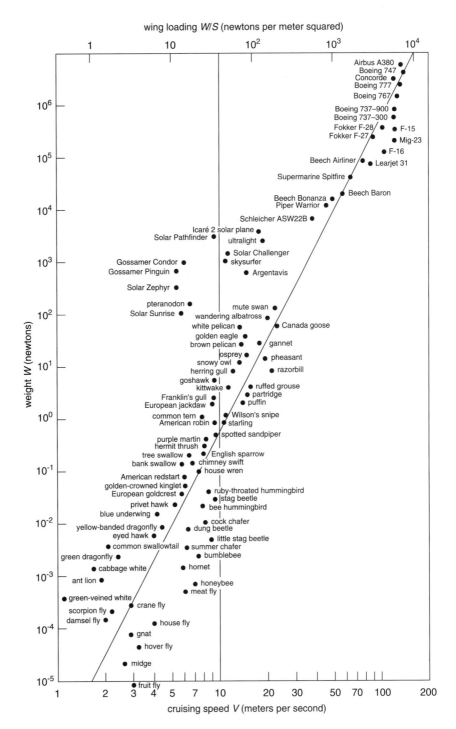

Figure 2 The Great Flight Diagram. The scale for cruising speed (horizontal axis) is based on equation 2. The vertical line represents 10 meters per second (22 miles per hour).

Very few phenomena in nature cover so wide a range; the Hertz-sprung-Russell diagram in astronomy is the only other one I am aware of. At the very bottom of the graph we find the common fruit fly, *Drosophila melanogaster*, weighing no more than 7×10^{-6} newton (less than a grain of sugar) and having a wing area of just over 2 square millimeters. At the top is the Boeing 747, weighing 3.5×10^6 newtons, 500 billion times as much as a fruit fly. The 747's wings, with an area of 511 square meters, are 250 million times as large. Despite these enormous differences, a 747 flies only 200 times as fast as a fruit fly.

Allow yourself time to study figure 2 carefully. It is loaded with information. The ascending diagonal running from bottom left to top right is the scale relation of equation 3. The constant c has been set equal to 47, almost twice as large as the value in figure 1. The vertical line marks a cruising speed of 10 meters per second, corresponding to 22 miles per hour and to force 5 on the Beaufort scale used by sailors and marine meteorologists. Birds that fly slower than this (those to the left of the vertical line) may not be able to return to their nest in a strong wind. (To return home in a headwind, a bird must be able to fly faster than the rate at which the wind sets it back.)

Deviations from the rule can be seen both to the left and to the right of the diagonal representing the scale relation of equation 3. The diagonal acts as a reference, a "trend line," a standard against which individual designs can be evaluated. Let's start with the birds and airplanes that follow the trend—the commonplace types found on or near the diagonal. The starling is a good example. A thrush-size European blackbird, 100 of which were released in 1890 in New York's Central Park, it has become a most successful immigrant (and somewhat of a nuisance, too). With a weight of 0.8 newton (80 grams, a little over 3 ounces) and a wing loading of 40 newtons per square meter, the starling is clearly an ordinary bird and does not have to meet any special performance criteria. But the Boeing 747 also follows the trend. In its weight class the 747 is a perfectly ordinary "bird," with ordinary wings and a middle-of-the-road wing loading. The weight of the 747 is no longer very special, either: today several other planes of similar weight are in service.

Deviations from the trend line may be necessary when special requirements are included in the design specifications. The 747's little brother, the 737, weighs only 50 tons (5×10^5 newtons), one-eight the weight of a 747-400. If the 737 had been designed as a scale model of the 747, its wing loading would have been half that of the larger plane (the cube root of 8 is 2). And according to equation 2 its cruising speed would have been only 71 percent of its big brother's: not 560 miles per hour but only 400. This would have been a real problem in the dense air traffic above Europe and North America, where backups are much easier to avoid if all planes fly at approximately the same speed. To make it almost as fast as the 747, the 737 was given undersize wings. Its wing loading is higher than those of ordinary planes of the same weight class, and it is therefore located to the right of the trend line in figure 2. (With a cruising speed 60 miles per hour less than that of the 747, the 737 would still be a bit of a nuisance in dense traffic were it not consigned to lower flight levels.)

Far left of the diagonal, in the center of figure 2, is *Pteranodon*, the largest of the flying reptiles that lived in the Cretaceous era. Weighing 170 newtons (37 pounds), it was almost twice as heavy as a mute swan or a California condor. It had a wingspan of 23 feet (7 meters) and a wing area of 108 square feet (10 square meters)—comparable to a sailplane. Its wing loading was 17 newtons per square meter, about one-tenth that of a swan but comparable to that of a swallow. *Pteranodon* spent its life soaring above the cliffs along the shoreline, since its flight muscles were not nearly strong enough for continuous flapping flight. Its airspeed was about 7 meters per second (16 miles per hour)—not fast for an airborne animal that must return to its roost in a maritime climate. However, there were no polar ice caps during the Cretaceous era, and there was less of a temperature difference between the equator and the poles than there is today; as a result there was much less wind.

The largest flying animal that ever lived, however, was not a reptile, but a giant bird that roamed the windy slopes of the Andes and the pampas of Argentina 6 million years ago. Looking much like an oversized California condor, *Argentavis magnificens* weighed 700 newtons (150 pounds). With a wing area of 8 square meters and a wingspan of 7 meters, its speed in soaring flight was about 15 meters per second, much the same as that of a golden eagle. It defies the conventional wisdom that birds much heavier than 25 pounds cannot fly. Exceptions to the rule add spice to the work of a scientist.

After centuries of experimentation, humans finally managed to fly under their own power. That required feather-light machines with extremely large wings. The only way to reduce the power requirement to a level that humans could attain was to reduce the airspeed to an absolute minimum. Humans pedaling through the air on gossamer wings are the real mavericks in the Great Flight Diagram (figure 2). They are represented there by Paul McCready's *Gossamer Condor*, the first successful example of the breed. Also shown are a number of solar-powered planes. A severe lack of engine power forces them also to the far left of the trend line. As a mode of transportation they are just as fragile as human-powered planes or extinct flying reptiles. We'll have to wait for much more efficient solar cells before solar-powered flight will succeed in the struggle for survival in this technological niche.

What about the Concorde? Wasn't it supposed to fly at about 1,300 miles per hour? How come it didn't have higher wing loading and therefore smaller wings? The answer is that the Concorde suffered from conflicting design specifications. Small wings suffice at high speeds, but large wings are needed for taking off and landing at speeds comparable to those of other airliners. If it could not match the landing speed of other airliners, the Concorde would have needed special, longer runways. The plane's predicament was that it has to drag oversize wings along when cruising in the stratosphere at twice the speed of sound. It could compensate somewhat for that handicap by flying extremely high, at 58,000 feet. Still, its fuel consumption was outrageous.

Convergence and Divergence

The Great Flight diagram (figure 2) exhibits many curious features. Let me name a few. Sports planes tend to be underpowered, but crawl toward the trend line as the engine power increases. Small birds fly much faster than computed when they are migrating. Insects are either too fast for their size or too slow. Large soaring birds deviate more from the trend line than their smaller cousins. Large airliners tend to have the same wing loading, irrespective of size. Biologists believe that creatures that exhibit better all-around performance have a better chance to survive. They tend to evolve in similar ways, much as insects, birds, and airplanes cluster around the trend line in figure 2. The label given to this idea is *convergence*. In short, evolutionary success is determined by functional superiority. Good designs perform better than alternative ones, so alternative solutions are weeded out. Creatures that venture far from the trend line, human-powered airplanes for example, have little chance of survival in the long run. In fact, human-powered airplanes have become extinct.

In the very beginning of powered flight, airplanes tended to be underpowered. Early aircraft engines weighed many pounds per horsepower. In order to keep the total weight within limits, relatively small engines had to be used. One hundred years ago, the cruising speed of most airplanes was 40 miles per hour at best. The advantage is obvious: those early planes could take off and land on grass strips. Also, crashes were relatively easy to repair. The major disadvantage was that these planes had to be kept on the ground in high winds. The best fighters in World War I were a lot faster: with engines delivering up to 200 horsepower, speeds of 100 miles per hour or more could be obtained. In World War II, Mustangs and Spitfires reached speeds up to 450 miles per hour.

When you decide to install a more powerful engine in your next plane, the total weight will increase because the engine is heavier and the fuel tank bigger. This requires a larger wing. But a larger engine allows you to fly faster, and that permits you to choose a smaller wing. The net result is that the wing area stays about the same as engine power increases. Typical private planes have a wing area of about 20 square meters. When S is fixed, the wing loading W/S does not increase as the third root of W; it increases in linear proportion to W itself. This is exactly what happens as you move from the Wright Flyer or the Skysurfer to the Beechcraft Baron and the Beechcraft Bonanza in figure 2. A clear case of evolutionary convergence: as aircraft engines improve in terms of horsepower per pound of engine weight, it pays to install a larger engine, which allows a higher cruising speed. The trend line is rejoined at speeds around 60 meters per second (130 miles per hour), a typical plane then weighing about 4,000 pounds. This is just one example of the rapid pace of convergence in technological evolution.

Curiously, the Supermarine Spitfire, the famous British World War II fighter, is right on the trend line in figure 2. Thus, you might think it is rather ordinary. But sometimes appearances are deceiving. With a wing area of 22.5 square meters and a takeoff weight of 40 kilonewtons, a Spitfire's cruising speed computes as 69 meters per second (250 kilometers per hour, 155 miles per hour). What about the reported top speed of 700 kilometers per hour, then? And why was a 1,600-horsepower Rolls-Royce Merlin engine installed? Spitfires were interceptors: they had to climb to 25,000 feet just in time to attack approaching German bombers. That is what the famous 48-valve Merlin engine was for. You can't fly fast and climb fast at the same time. It pays to have a rather low cruising speed, because most of the power then can be used to climb fast. If you plan on modifying a Spitfire for racing, you should give it much smaller wings and forget about a high rate of climb. Taking off from grass strips then also is out of the question.

Why doesn't the Great Flight Diagram (figure 2) include any bats? The diagram is terribly crowded as is. Also, no new information would have been presented. Bats' wing loading is similar to that of birds of the same size. By omission, the case for convergence is made stronger yet: having to live in the same environments, birds and bats have evolved toward comparable aerodynamic design parameters. There are subtle differences, though. The largest swan weighs about 25 pounds, but the largest bat only 5 pounds. This is probably not a matter of muscle power but a consequence of lung design. The lungs of birds have air sacs behind them, so they are ventilated twice during each respiration cycle and can pick up much more oxygen than the lungs of mammals, which don't have those lung extensions. I wonder why evolution hasn't solved this discrepancy. Is the cause of the handicap that mammals appeared on the scene so much later than birds?

Sometimes a limited amount of divergence from the trend line is unavoidable. Vultures and eagles prefer to soar in "thermals" (ascending pockets of hot air) and need a rate of descent less than 1 meter per second in still air. Since these large birds can glide 15 meters for every meter of altitude lost, they should not fly faster than 15 meters per second (33 miles per hour). The wing loading of large soaring birds therefore is fixed at about 80 newtons per square meter. The extinct giant Andes condor *Argentavis magnificens* is no exception. As they grow bigger, the large soaring birds diverge further from the trend line in figure 2. Their lifestyle requires much less muscle power than those of geese and swans, so their flight muscles are relatively small. Continuous flapping is out of the question; they have wait until sufficiently strong thermals develop in the course of the day. It shouldn't surprise you that smaller species start their soaring days earlier than larger ones. Neither should it surprise you that a flock of soaring birds sends scouts aloft in the morning in order to find out whether the updrafts have become strong enough.

Large birds that cannot soar but have to flap their wings have problems of their own. As far as wing loading and flight speed are concerned, swans, geese, and ducks follow the trend line faithfully. But they don't grow much heavier than about 25 pounds. So where's the rub? The muscle power available to flapping birds is about 20 watts per kilogram of body weight. Muscle power is proportional to weight, but the power required to maintain horizontal flight is proportional to the product of weight and flight speed. Bigger birds have to fly faster, so their power reserve decreases as their weight grows larger. The largest species of swans have very little power to spare. According to Swedish researchers, they can gain altitude no faster than 50 feet per minute. I can confirm this number from personal experience. One autumn many years ago, a flock of mute swans landed in a meadow behind my house. After resting for a day and filling their stomachs, they took off. The meadow, however, was surrounded by brushwood and trees on all sides. The leader of the flock realized that he couldn't clear those obstacles head-on and decided instead to fly a large circle, exploiting the width of the meadow. Slowly the flock gained height. After a circle and a half, they cleared the brush on the southwest corner of the field.

From swans and eagles to insects: a large step down in weight, but similar characteristics of convergence and divergence. Big birds either soar slowly with oversized wings or follow the trend line by working hard and flying fast. Among insects, the slow ones, butterflies and the like, follow the trend line rather faithfully, but shifted a bit to the left. Many butterflies are capable of gliding and soaring, and use these styles of flight to conserve energy. If they can take advantage of strong tailwinds, migrating monarch butterflies cross the Gulf of Mexico directly, instead of following the coast. They have been observed by radar to flock in the updrafts between the "roll vortices" in the lower atmosphere that stretch at a slight angle to the wind direction. They float without flapping—a perfect way to cross 500 miles of open sea. I don't know how they find out where the updrafts are, but I do know how human observers do : under appropriate circumstances, "cloud streets" form in the air between each pair of roll vortices.

Mosquitoes, bees, and flies fly in an entirely different way. Their buzzing wings act like the rotor blades of helicopters. Their wing size is not determined by their flight speeds but by their flapping frequency. The speeds suggested in figure 2 are therefore not reliable. Bees can go faster when they have to; 10 meters per second is not uncommon. Some biologists argue that bees diverge farther from the trend line the smaller they become. On the other hand,

Storm petrel (*Hydrobates pelagicus*): $W = 0.17$ N, $S = 0.01$ m^2, $b = 0.33$ m.

the very smallest flies, such as *Drosophila melanogaster*, go only 1 meter per second, one-third as fast as figure 2 suggests. Going into the details of the performance of buzzing insects would lead me astray. For very small creatures, air does not obey the aerodynamic principles that are valid for birds and planes. To a fruit fly, for example, flying must feel very much like swimming in syrup. (For those who want to know more, I recommend reading one of the books on insect flight listed in the bibliography. For most readers, David Alexander's *Nature's Flyers*, though not limited to insects, is by far the best source. And those who insist learning about all the intricate details will have to study Robert Dudley's book *Biomechanics of Insect Flight*.)

In the top right corner of figure 2, another constraint occurs. It is the speed of sound. For good reasons, explained in chapter 6, airliners travel above 30,000 feet, where the speed of sound is 295 meters per second (1,062 kilometers per hour, 664 miles per hour). They have to fly somewhat slower than that, typically 560 miles per hour, in order to avoid making little shock waves that increase airframe drag rapidly as the speed of sound is approached. Curi-

Barn swallow (*Hirunda rustica*): $W = 0.2$ N, $S = 0.013$ m^2, $b = 0.33$ m.

ously, all modern long-distance planes cluster around the original design parameters of the Boeing 747. Convergence in this case is not just toward the trend line but to a quite specific weight class: a small cloud of data points in the top right corner of figure 2. The Airbus A380 is no exception. (Chapter 6 deals with the engineering logic that has led to this curious development. I did explain the logic in the first edition, but I did not see the consequences for the size of future airliners at the time.)

Incidentally, the Boeing 747 is represented in figure 2 as having a wing loading of 7,000 newtons per square meter and a cruising speed of 136 meters per second. But 136 meters per second is 300 miles per hour, roughly half the 747's actual cruising speed. What has gone wrong here? In figure 2 the lower air density at cruising altitudes has been ignored. Since the air density at 39,000 feet is only one-fourth the density at sea level, the high-altitude cruising speed is twice the cruising speed near Earth's surface. Figure 2

Bee hummingbird (*Mellisuga helenae*): $W = 0.02$ N, $S = 0.0007$ m^2, $b = 0.07$ m.

gives the speed at sea level; table 6 (in chapter 6) gives the necessary conversion factors.

Traveling Birds

Several groups of ornithologists have been making radar measurements of actual flight speeds of migrating birds. The Schweizerische Vogelwarte (Swiss Ornithological Institute) published a massive list of radar speed data in 2002, and biologists at the Flight Ecology Department of the University of Lund in Sweden added their own list in 2007. (A selection of these data is presented in the appendix.) Theoreticians have begun to dissect the assumptions underlying the "aerodynamic theory of bird flight," the theory from which I distilled my way of handling the matter.

Since 1970 or thereabouts, everyone involved with bird flight assumed that the speed at which birds glide best is not too different from the most economical speed in flapping flight. We now know this was an unwarranted simplification. If flapping birds want to conserve energy, they have to fly much faster than when gliding. When birds are in no hurry, like a circling flock of homing pigeons or a great dancing swarm of starlings at sunset, they fly at a speed that requires the least muscle exertion. It turns out that this speed is not too different from the optimum gliding speed. But bird migration is another business altogether. Birds on migration often are in a hurry. Most of them fly faster than the speed that minimizes their "fuel consumption" per hour, near the top speed their muscles allow. The migration speed of small birds may be as much as twice the speed that requires least muscle power. In fact, I have found no numbers below 10 meters per second (22 miles per hour) for any songbird on migration. If you want to amend figure 2, here is your chance. All the data you need are given in the appendix. A typical 10-gram songbird migrates at 10 meters per second, 80-gram starlings and half-pound jackdaws manage 14 meters per second (30 miles per hour), and many wading birds fly 20 meters per second (45 miles per hour) when they are making their seasonal long-distance journeys. If you want to make a sophisticated correction to figure 2, you should choose a curve that makes the flight speed much less dependent on weight instead of the trend suggested in figure 2. That would account for the fact that small birds have lots of power to spare for speeding, while large birds are straining themselves even when flying most economically.

Flapping, Gliding, Soaring, and Landing

What about the various swifts, swallows, and martins in figure 2? They are all found on the left of the trend line. For their weight, they all have rather large wings and fly relatively slowly. There must be something wrong here. Swifts did not get their name for nothing.

In fact, swifts and swallows are fast only when gliding, diving, or fooling around. In level flapping flight, they are not fast at all. Radar data on migrating swifts give speeds around 10 meters per second (22 miles per hour). In wind tunnels, swallows fly no faster than 12 meters per second (27 miles per hour). Their comfortable cruising speeds are lower yet, consistent with the data in figure 2.

Swifts and their relatives can fly very slowly, when they have to, by spreading their wings wide. When they want to fly faster, they

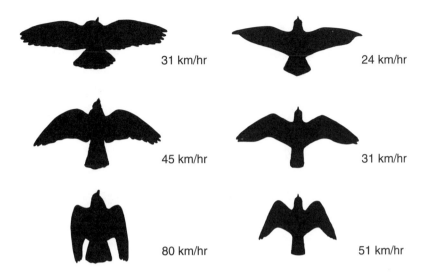

31 km/hr   24 km/hr

45 km/hr   31 km/hr

80 km/hr   51 km/hr

Figure 3 Birds progressively fold their wings as their speed increases. On the left is a pigeon, on the right a falcon. At high speeds, fully spread wings generate unnecessary drag. This can be avoided by reducing the wing area.

can fold their wings. The elegance of their streamlining does not suffer when they reduce their wing area, but the wing loading increases, and with it the cruising speed. Are they poking fun at the laws of nature? According to equation 2, a bird cannot alter its speed at will if it wants to fly economically, once blessed with a particular set of wings. The cruising speed is controlled by the wing loading: $W/S = 0.38V^2$. But if S can be changed to fit the circumstances, this problem vanishes: the cruising speed then changes too. All birds do this to some extent, though not always with the grace and sophistication of swifts and swallows. But gliding, soaring, and maneuvering are altogether different from flapping. In the downstroke of flapping flight, all birds spread their wings fully; however, when gliding, birds can fold their wings at will. Figure 3 shows how gliding falcons and pigeons progressively fold their wings as their speed increases.

When low speeds are needed, all birds make their wing area as large as is possible. The sparrow hawk on final approach (figure 4)

Figure 4 Sparrow hawk (*Accipiter nisus*): $W = 2.5$ N, $S = 0.08$ m^2, $b = 0.75$ m.

is a good example. Since it wants to fly slowly, it spreads its primary quills and tail feathers wide. In fact, many birds deliberately stall their fully stretched wings on final approach, maximizing drag to obtain quick deceleration and not caring about lift during the last seconds of flight. Just for fun, watch pigeons landing on a tree branch or a rooftop, and see how they do it. Airplanes fully extend various slats and flaps in preparation for landing. Airplanes and birds alike minimize their landing speed to reduce the length of runway required or the risk of stumbling.

Swifts' amazing aerial maneuvers are made possible by the superb aerodynamic performance of their sweptback wings. I have seen them dallying in the updrafts in front of the cliffs in southern Portugal, first diving toward the shore and then suddenly shooting up like rockets and disappearing out of sight. In these stunts, flapping would be of no use. With their wings folded far back, swifts have another surprise in store. If they have to make a quick maneuver, they can generate a "leading-edge vortex" over their swept-back wings by suddenly increasing their angle of attack. Almost but not quite stalling their wings, they achieve a large momentary increase in lift that way, which allows for very sharp turns. This is how they catch insects in their flight path, and this is how they show off during the sophisticated aerobatics of courtship.

Continuous flapping flight does not support such extreme behavior. Level flapping flight is boxed in by a large number of constraints—kinematic, dynamic, energetic, physiological, and so on. When flapping, wings have to act not only as lift-generating surfaces but also as propellers, a combination never successfully imitated by human technology. Wings act as propellers only in the downstroke. The upstroke is of little use. Many birds fold their wings before they start the upstroke; others drastically reduce the angle of attack before their wings move upward. To keep things simple, I will assume that only the downstroke counts. This means that flapping wings are useless during one half of each wingbeat cycle, and have to produce twice the lift during half the time in order to make sure a bird stays airborne. The immediate consequence is that birds have to endure a roller-coaster ride when flapping at speed. This is obvious when you watch traveling geese fly by overhead. Their heads are kept steady, presumably to make sure that their delicate navigation equipment is not affected, but their bodies are shaking up and down with each wingbeat. Another consequence of the two-stroke cycle of flapping wings is that the angle of attack during the downstroke has to be much larger than when gliding at the same speed. This is fine as far as the flight muscles are concerned, because the airspeed for most economical gliding does not differ much from the speed that requires the least muscle effort when flapping. (Just watch any bird switching from gliding to flapping or vice versa, without change in speed.) But it does pay to choose a higher airspeed in flapping flight, because a bird can also

Cockchafer (*Melolontha vulgaris*): $W = 0.01$ N, $S = 0.0004$ m^2, $b = 0.06$ m.

get twice the lift by flying 40 percent faster (the lift goes as the square of the speed, and the square root of 2 is about 1.4). This option keeps the angle of attack at a value that doesn't compromise the aerodynamic performance of the wings. I know I am not doing justice to the great variety of flapping styles that birds employ, but a useful rule of thumb is that the most economical speed for flapping is 40 percent higher than that for gliding, provided a bird has no shortage of muscle power. Swans and other big birds do not have that option; their speed is limited by their muscle power. This implies that their wings are working at a high angle of attack during the downstroke, an angle that compromises flight efficiency somewhat. The whistling noise made by the flight feathers of mute swans proves that in the downstroke their wings are almost stalling.

Birds and Insects

A curious feature of figure 2 is the continuity between the largest insects and the smallest birds. The largest of the European beetles, the stag beetle *Lucanus cervullus*, weighs 3 grams, about the same as a sugar cube or a fat hazelnut. The smallest bird on Earth, the Cuban bee hummingbird *Mellisuga helenae*, weighs 2 grams. The smallest European bird, the goldcrest, weighs 4 grams. Small bats also weigh about 5 grams, notwithstanding their different flight apparatus. The wing loadings of large insects do not differ much from

Stag beetle (*Lucanus cervus*).

those of small birds, either. This is no minor observation. In theory, conditions may be imagined in which the largest beetle exceeds the smallest bird in size, or a wide gap exists between the largest flying insects and the smallest birds. Such a gap does exist between the largest birds and the smallest airplanes, after all. And there are substantial construction differences, too. The exoskeletons of insects are made up of load-bearing skin panels, while birds (like humans) have endoskeletons, with the load-bearing bones inside the body. Notwithstanding the different construction techniques, the transition from insects to birds is barely perceptible. Apparently, the choice between an exoskeleton and an endoskeleton is a tossup for weights around 3 grams. Just a little heavier and the exoskeleton loses out to the little birds; just a little lighter and the endoskeleton has to make way for the big beetles. What factor determines this switchover? Is it the wing structure, the weight of the skeleton, the geometry of the muscle attachment points, the respiration constraints, or the blood circulation? This would be a wonderful research project for a young aeronautical engineer. Some experience with aircraft construction would give the engineer a head start. Like insects, most airplanes have exoskeletons:

their skins carry most of the structural load. For very small or very large flying objects, an endoskeleton is apparently not a wise choice.

The time has come to look deeper into the energy required for powered flight. Hummingbirds and jetliners consume a few percent of their body weight in fuel per hour. That is a sure sign that energy consumption is a major consideration in flight performance. Most of the time, flying is hard work. When you hear a wren sing its staccato "tea-kettle, tea-kettle, tea" in your backyard, it is not enjoying an idle moment; it is trying to keep competitors off its territory without having to patrol the perimeter. Flying back and forth would use up too much energy. Birds that have to spend much of the day looking for food find it easier to whistle a tune than to chase intruders. Similarly, birds feeding their nestlings must select their food carefully, choosing between the fattened caterpillars in the woods a quarter-mile away and the starving maggots in the meadow below. If a bird doesn't take care, it will spend more energy on getting food than it and its young get out of it.

Sparrow hawk (*Accipiter nisus*): $W = 2.5$ N, $S = 0.08$ m^2, $b = 0.75$ m.

An insurance company physician may put you on a fancy exercise machine, turn up the dial to 150 watts, and wait impassively for you to lose your breath. Having achieved that, the physician may turn the dial back a little to determine the load at which your heart stabilizes at 120 beats a minute. The Consumers Union does much the same with cars, driving them hard on a test bench to measure engine power, fuel consumption, and emissions. Aeronautical engineers do their testing in wind tunnels, where powerful fans simulate the high speeds needed to put scale-model airplanes through their paces. Aircraft companies need to know the flying properties of their new planes in great detail before they send someone off on a test flight.

Wind tunnels come in many shapes and sizes. The simplest is a straight pipe with an adjustable ventilator at one end, but most are much more elaborate. The supersonic wind tunnels used to test models of jet fighters and of the Space Shuttle require enormous amounts of power. The very largest have working cross-sections several meters wide and several meters high. Wind tunnels are used not only to test airplanes but also to investigate how air flows around full-scale cars and even to perform environmental impact studies of industrial installations.

My favorite example of wind-tunnel work dates from 1968. Vance Tucker, a zoologist at Duke University, trained a budgerigar (grass parakeet, *Melopsittacus undulatus*) to perform flapping flight in a specially built wind tunnel. In order to measure its oxygen consumption, he fitted the little bird with an oxygen mask. That way he could get the data he needed to calculate how much energy the bird used at various airspeeds (ranging from 10 to 30

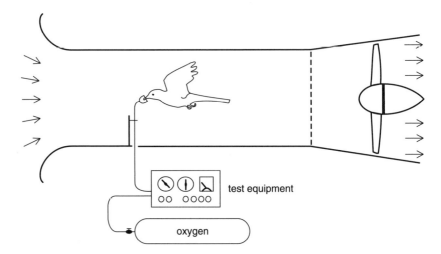

Figure 5   Vance Tucker's parakeet in the wind tunnel.

miles per hour) under various conditions, in horizontal flight and during 5° ascents and descents (figure 5).

To convert his oxygen-consumption data into useful form, Tucker first had to calculate the energy consumed during each flight. Next he had to subtract the energy spent maintaining the bird's metabolism (which would not be available for flight propulsion). The basal metabolic rate of small birds is roughly 20 watts per kilogram of body weight, 10 times the rate in humans. Tucker's parakeet weighed 35 grams, so its body used 0.7 watt to sustain itself. Finally, since the efficiency of the conversion from metabolic to mechanical energy is only about 25 percent, the net propulsive power is one-fourth the metabolic cost of flying. The net mechanical power of the flight muscles is plotted in figure 6.

The most striking feature of figure 6 is that slow flight is uneconomical. It is easy enough to understand that the faster you travel the more power you need; riding a bicycle or driving a car will have taught you that. But birds also need a lot of power to fly slowly. For this reason, the power required for flapping flight has a minimum in the middle of the speed range. (See figure 6.) In hor-

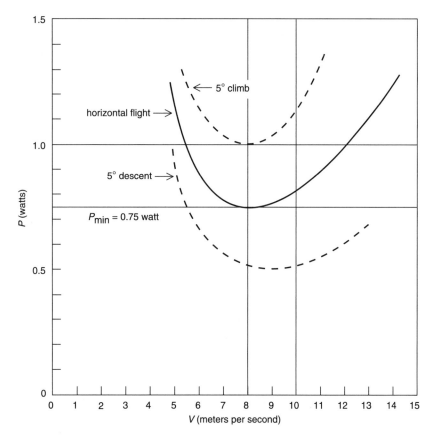

Figure 6 The flight performance of Vance Tucker's parakeet. In horizontal flight the most economical conditions were obtained at about 8 meters per second, with the bird supplying 0.75 watt of mechanical power. In climbing flight more power is needed, in descending flight less.

izontal flight, the most economical speed for Tucker's parakeet was slightly more than 8 meters per second (18 miles per hour). At that speed it required 0.75 watt, equivalent to 0.001 horsepower, to remain airborne. At lower and higher speeds, more power was needed.

Low-speed flight is uneconomical because birds and airplanes have to push the air surrounding them downward in order to stay airborne. Relatively little air flows around the wings when the airspeed is low. To sustain its own weight, a bird or an airplane must give that small quantity of air a powerful impulse. That requires a lot of energy. When a bird is flying fast, though, a lot of air crosses its wings; only a little push is needed to remain airborne then, and that requires much less energy. (This will be taken up again in chapter 4.)

Tucker's parakeet apparently could not fly slower than 5 meters per second (11 miles per hour). At that speed, flapping its wings at full power, its flight muscles generated about 1.3 watt, evidently its maximum continuous power output. Let's do a few calculations to see if these numbers make sense. The pectoral muscles of a bird account for about 20 percent of its body weight; in continuous flapping flight, a bird can generate about 100 watts of mechanical power per kilogram of muscle mass. A human can manage only about 20 watts per kilogram of muscle mass.

Since the pectoral muscles of a 35-gram parakeet weigh about 7 grams, the corresponding continuous power output should be about 0.7 watt, corresponding nicely to the minimum in figure 6. The maximum continuous power of flight muscles is roughly twice the normal rate, or 200 watts per kilogram of muscle mass. For the parakeet this works out to 1.4 watt, pretty close to the experimental data in figure 6. When the rules of thumb you are using agree with the results of measurements, you know are on the right track.

Energy, Force, and Power

We have made a terrible mess of simple physical concepts in ordinary life. We treat force, power, and energy as if they were interchangeable. Straightforward, unequivocal definitions of these quantities have been around for 200 years, and yet we continue to be confused. The scientific definition of the word 'power' is "rate of doing work," nothing more and nothing less. In everyday language, of course, 'power' has many other connotations. My Random House Dictionary refers to capabilities and capacities, to political strength (incidentally, wouldn't "strength" be a force?), to control and command, and so forth. Half a column goes by before the dictionary gets around to "in physics, the time rate of doing work." Indeed, the majority of concepts associated with the word 'power' tend to confuse rather than enlighten. The same holds for 'force' and 'energy'. The Danes speak of nuclear "force" when they mean nuclear energy. (Some Danish bumper stickers used to proclaim "Atomkraft—nej tak," meaning "Atomic energy—no, thanks.") The Dutch and the Germans use the equivalent of 'horse-force' instead of 'horsepower'. Psychologists, whatever their native tongue, refer to mental energy as though it satisfied some mechanistic conservation law.

When we are trying to describe nature, we need clear and unambiguous definitions. Let us agree, then, that a force is the intensity with which I push or pull at an object, whether it starts to move or not. Exerting as much force as I please, I perform no work unless the object is displaced in the same direction. Sideways displacements do not count. Work is performed in proportion to the distance the object is displaced. Work is a form of energy, and the time rate at which it is consumed or supplied is called power.

When calculating the amount of work performed, we must consider both the force applied and the distance the object has moved. What is more straightforward than that work equals force times distance? That indeed is the accepted definition. The corresponding definition of 'power' is equally straightforward: power is a rate of doing work, or a certain amount of energy expended per second. Since work equals force times distance, power must be force times distance per second. But distance traveled per second is what we call speed. Therefore, power equals force times speed.

Before continuing with the flight performance of birds, we must define the proper units for work and power. In chapter 1 we settled on the newton (102 grams) as our force unit. Defining the units for work and power is simple. Work equals force times distance, so in the metric system it must be calculated in newton-meters. This is a unit in itself, called the *joule* after the British physicist James Joule (1818–1889), who performed a brilliant energy-conversion

Sanderling (*Calidris alba*).

experiment in 1845. Power equals energy per second and so must be computed in joules per second. But because a joule equals a newton-meter, power may also be calculated in newton-meters per second; that comes to the same thing. The unit for power also has a name of its own: the *watt*, after James Watt (1736–1819), the Scottish inventor of the steam engine. We already used watts when discussing the power requirements of Tucker's parakeet. (James Watt used another power unit, for which he coined the word 'horsepower'. He needed a word that would kindle the imagination of coal mine directors who were in the market for replacing their pit ponies with one of his steam engines. One horsepower is equivalent to about 750 watts. Because there are 1,000 watts in a kilowatt, a horsepower equals 0.75 kilowatt.)

The quickest way to get used to a new set of concepts is to play with them. You know from experience that it takes energy to walk uphill, because you have to lift your own weight against the pull of gravity. How much energy does it take to climb one flight of stairs? Let's assume that you weigh 70 kilograms (about 154 pounds, or 700 newtons) and that the vertical distance between floors is 3 meters (about 10 feet). Since energy equals force times distance, it takes 2,100 joules to climb from floor to floor. Is that a lot? Not at all. A gram of peanut butter contains 27 kilojoules (27,000 joules) of nutritional energy. If your body converts 20 percent of that to useful work, it will have 5,400 joules to expend. This means that you can climb more than two flights of stairs on a single gram of peanut butter. Walking up and down flights of stairs at the office is healthy exercise, but not an efficient way to lose weight.

The *power* you need to walk up a flight of stairs (that is, the *rate* at which energy has to be supplied) is not negligible, however. If your vertical speed on the stairs is 0.5 meter per second, only half the speed typical of a leisurely hiking trip, the climbing power you need (force times speed) is 700 newtons $\times$ 0.5 meter per second, which equals 350 watts. Only a professional athlete can maintain such a rate for more than a minute or so. A healthy amateur can maintain a power output of 200 watts for less than an hour; a professional athlete can maintain that rate for several hours.

Lance Armstrong and Other Athletes

Running up a flight of stairs is hard work, but so is bicycle racing, and so is flying. In one-hour runs, Lance Armstrong managed to deliver 500 watts of mechanical power to the crankshaft of his bike. At a muscle efficiency of 25 percent, this amounts to about 2,000 watts of metabolic power, or 20 times Armstrong's metabolic rate at rest. His speed over the one-hour run was 14 meters per second (32 miles per hour), and his heart beat about 190 times per minute, a little over 3 beats per second. Armstrong carefully synchronized his heart and his legs: he cranked 95 rpm, exactly half his heart rate. Very few people can maintain such rates for any length of time. Assuming that Armstrong's heart rate at rest is about 40 beats per minute (the big heart of a professional athlete beats slower than the hearts of ordinary people), he could apparently maintain a frequency almost 5 times higher than at rest.

How do Armstrong's numbers compare with those of avian athletes? Barnacle geese (*Branta leucopsis*) equipped with miniaturized instruments were extensively tested in flights behind a speeding truck on an abandoned runway. These are big birds, with a weight of 16 newtons (more than 3 pounds), a wingspan of 1.35 meter, and a wing area of 0.17 square meter (almost 2 square feet). Their heart rate at rest was 72 beats per minute, not much higher than that of 140-pound humans. When these geese were shuffling around in anticipation of a flying session, their heartbeat frequency doubled to 150 beats per minute. And when they were gliding at a speed of 14 meters per second (as fast as Armstrong) it went up to 260 beats per minute. But flapping their wings and speeding behind the truck at an amazing 19 meters per second (43 miles per hour, much faster than Armstrong could maintain for any length of time on level terrain), these geese had a heartbeat frequency of 510 times per minute, fully *seven* times the rate at rest. At that speed they produced an estimated 25 watts of mechanical power. At a muscle efficiency of 25 percent, this corresponds to 100 watts of metabolic power, only 14 times their metabolic rate at rest. Clearly, Lance Armstrong outperformed the geese on this score. The "metabolic scope" of humans, dogs, and horses is larger than that of birds. Even though Armstrong produced just 7 watts per kilogram, he is the real star athlete here. He can't help it that humans, being mammals, have lungs that are poorly ventilated relative to those of birds. Bar-tailed godwits make week-long nonstop flights at *9 times* their basal metabolic rate.

While we are at it, let's have a look at the impression of speed that flying birds generate. When I stroll around a shopping mall, I walk about half a meter per heartbeat. A little sparrow flying from rooftop to rooftop goes about as slow: with a speed of 7 meters per second and its heart beating 16 times per second, it covers less than half a meter per heartbeat. At the other extreme, when Lance Armstrong races at top speed he proceeds more than 4 meters per

Ruby-throated hummingbird (*Archilochus colubris*): $W = 0.03$ N, $S = 0.0012$ m^2, $b = 0.09$ m.

heartbeat. A barnacle goose flies only 2 meters per heartbeat, which isn't much when one considers the high speed.

Back to Tucker's Parakeet

The performance of Tucker's parakeet was measured both in horizontal flight and in ascents of 5° (which corresponds to a slope of 8.7 percent). Flying at the minimum-power speed of 8 meters per second, the parakeet achieved a rate of ascent of 0.7 meter per second. Now let us repeat the calculation we did a minute ago. Power equals force times speed. The force we are talking about here is the force needed to lift the bird's weight upward. The parakeet weighed 0.35 newton (35 grams, a little more than an ounce), and the climbing power required was 0.35 newton × 0.7 meter per second, which is about 0.25 watt. The power needed to sustain horizontal flight at a speed of 8 meters per second was 0.75 watt. In climbing flight, an additional 0.25 watt was needed, for a total of 1 watt. Figure 6 confirms this calculation.

One-fourth of a watt is not even enough to light a single bulb on a Christmas tree. A 350-ton Boeing 747's rate of climb immediately

Herring gulls (*Larus argentatus*): $W = 11.4$ N, $S = 0.2$ m^2, $b = 1.34$ m.

after takeoff is about 15 meters per second, or 3,000 feet per minute in aviation parlance. Quite apart from the power needed to remain airborne, the four jet engines of a 747 then must supply 50 million watts. That's equivalent to 17 top-of-the-line diesel locomotives producing 3,000 kilowatts (4,000 horsepower) each.

When a bird climbs it must exert itself to overcome the force of gravity, but when it descends gravity does some of the work. To descend at a glide slope of 5° at 8 meters per second, Tucker's parakeet needed only 0.5 watt of power: 0.75 watt to remain airborne, minus the 0.25 watt supplied courtesy of gravity. Taking this a bit further, we can easily work out the glide angle at which the little bird no longer needs to flap its wings: about 15°, corresponding to a glide slope of about 26 percent. Thus, for every meter of altitude it loses, the parakeet travels about 4 meters forward. A wandering albatross reaches 25 meters of forward travel for every meter of altitude lost, a gliding swift more than 10, an eagle 15.

Starling and Spitfire

Let's take a closer look at a starling. Its weight is 80 grams, so its flight muscles weigh no more than 20 grams. At full continuous power, the muscles produce 20 grams times 100 watts per kilogram (that is, 2 watts). How much power is needed? At a speed of 10 meters per second, I estimate that the aerodynamic drag is $\frac{1}{7}$ of the bird's weight, or 0.114 newton. Multiplying this by 10 meters per second, I find the power requirement to be 1.14 watt. Obviously, there is muscle power to spare at this speed, for example to gain altitude. With 0.86 watt remaining, a 0.8-newton bird can maintain a climb rate of a little more than 1 meter per second for some time—and much more if it is in a hurry. Roughly half of the power available is used for cruising, while the other half is on standby for climbing and maneuvering. A similar rule of thumb is valid for general-purpose sports planes. (I would not have dared to do these sums 15 years ago, because the sparse data on bird performance available then would have forced me to make a much higher estimate of the aerodynamic drag.)

Migration is a different matter altogether. If a migrating starling flies at a speed of 14 meters per second, as radar data indicate, its drag has increased to 0.133 newton ($\frac{1}{6}$ of its weight). Multiplying this number by the flight speed of 14 meters per second, I obtain 1.86 watt, only a little less than my estimate for the power available. No nitpicking is called for; these calculations are rough estimates, though I based them on wind-tunnel experiments with starlings in Flagstaff, Arizona. Some migrating birds fly with their throttle wide open, as it were. Small birds can fly substantially faster than the cruising speed suggested by figure 2.

We can do similar sums for the Supermarine Spitfire, the famous British fighter of World War II. The Spitfire Mark IX of 1944 had a Rolls-Royce Merlin 66 engine with an output of 1,700 horsepower, a loaded weight of 4,300 kilograms, a wingspan of 11.2 meters, and a wing area of 22.5 square meters. According to equation 2 in chapter 1, a plane with $W = 43,000$ newtons and $S = 22.5$ m^2 has an estimated cruising speed of 71 meters per second (255 kilometers per hour, or 160 miles per hour). In the thinner air at 25,000 feet, the cruising speed increases to 105 meters per second (236 miles per hour), well short of the Spitfire's reported top speed of 408 miles per hour.

How much power does a Spitfire need to cruise economically at 25,000 feet? A good guess for the drag at that speed is 10 percent of the weight—that is, 4,300 newtons (just short of 1,000 pounds). Drag times cruising speed gives cruising power; which computes as $4.3 \times 105 = 450$ kilowatts, or 600 horsepower. So what was the real use of the big 1,700-horsepower engine? The Spitfire was designed to climb to 20,000 feet within 6 minutes! Its initial rate of climb had to be 20 meters per second, which requires $20 \times 43 = 860$ kilowatts. That amounts to 1,150 horsepower, just for the climbing! Once up high, the spare power was used for dog-fights or high-speed chases. At 400 miles per hour, the drag has increased to about one-sixth of the weight; that requires the entire 1,700 horsepower available.

High Speed, Low Drag

Power equals force times speed. The power required in horizontal flight must be equal to some unknown force times the forward speed. But what is this force? It is the propulsive force, or thrust, T. To fly at a constant speed, a bird or a plane must develop just enough thrust to overcome aerodynamic drag D. (See chapter 4.) Since the two forces must be equal at constant speed, I'll talk only about the drag from now on. Drag times speed equals the power required, P, so drag equals power divided by speed. In mathematical symbols, these two concepts can be written as follows:

$$P = DV, \tag{4}$$

$$D = P/V. \tag{5}$$

It is quite simple to convert figure 6 into a graph depicting drag versus speed. For every point on the power curve for horizontal flight, we can determine P along the ordinate and V along the abscissa. Using equation 5, we can then compute the value of D at each speed. The curve obtained in this way is shown in figure 7, which demonstrates even more clearly than figure 6 that low-speed flight is extremely uneconomical. The minimum drag of a parakeet is

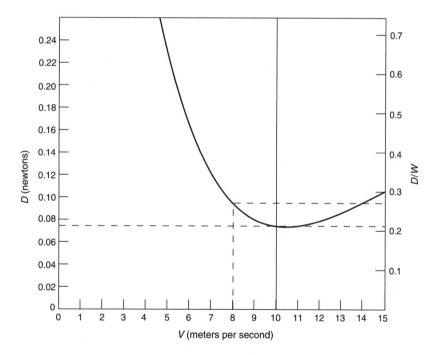

Figure 7   Aerodynamic drag, D, and ratio between drag and weight, D/W, for Tucker's parakeet. The minimum value of the drag is 0.078 newton (nearly 8 grams), and the lowest value of D/W is 0.22, both at a speed of nearly 11 meters per second (25 miles per hour). The speed at which the drag is smallest is higher than the speed at which the power has a minimum.

0.078 newton (almost 8 grams) at a speed of 11 meters per second (25 miles per hour), substantially above the speed at which the propulsive power reaches its lowest value. How does this happen? Power is the product of drag and speed (equation 4). Flying slowly is economical only if the drag doesn't increase too quickly when the speed comes down. The minimum power level obtains at a speed that compromises between a low value of D and a low value of V.

We must now decide whether figure 6 or figure 7 should determine our thinking about the energy consumption of birds and airplanes. In chapter 1 it was casually announced that an optimum cruising speed exists and that all calculations would be based on that optimum. Now comes this question: Should we use the lowest value of P, or the lowest value of D? The answer depends on what we wish to achieve. If we want to achieve the longest flight duration (as would be the case when an airplane is locked in a holding pattern while waiting for a landing slot), we must minimize energy consumption per unit time. Because power is energy per second, we can remain airborne longest at the lowest value of P. When time is the decisive factor, figure 6 is appropriate.

If P is the energy required per second, how much energy is consumed per meter? This is, of course, the quantity we want to minimize in order to maximize the distance we travel. The cruising speed we are looking for is the speed at which the energy consumed per meter of travel is as low as it can be. Table 2 is helpful here: energy equals force times distance, so force equals energy per unit distance. Forces can be measured in newtons, but joules per meter are just as good. In our case this means that the aerodynamic drag D is identical to the energy consumption per meter traveled. Flying at a speed that minimizes D, we have automatically minimized the energy consumed per unit distance. The cruising speed is the speed at which the smallest value of D is obtained. When distance is the decisive factor, figure 7 is appropriate.

Where does this lead us? The lowest energy consumption per unit distance is achieved at a relatively *high* speed. Birds and planes must fly *fast* to be economical. Imagine that cars were

Table 2 Various units for energy, power, force, and speed.

Energy

1 joule = 1 newton-meter = 1 watt-second

1 kilowatt-hour = 3,600,000 joules = 3.6 megajoules

1 calorie = 4.186 joule (1 joule = 0.239 calories)

1 kilocalorie = 4.186 kilojoules (used in nutrition)

1 megajoule = 0.278 kilowatt-hour = 239 kilocalories

1 British thermal unit (Btu) = 1,055 joules

Power

1 watt = 1 joule per second = 1 newton-meter per second

1 horsepower = 746 watts

Force

1 newton = 1 joule per meter

1 kilogram (force) = 9.81 newton

1 pound = 0.454 kilogram (weight) = 4.45 newtons

1 ounce = 28.4 grams (weight) = 0.278 newton

Speed

1 meter per second = 3.6 kilometers per hour

1 meter per second = 197 feet per minute

1 mile per hour = 0.44 meter per second

1 knot = 1 sea mile per hour = 0.506 meter per second

---

required to travel faster than 70 miles per hour to keep pollution and fuel consumption within limits, or that truckers were fined for going slower than 55! For that matter, just compare Tucker's parakeet to any land animal of comparable size. The cruising speed of the little bird was 25 miles per hour—well beyond the reach of mice, chipmunks, and squirrels. A rabbit can run for its life if it has to, and can reach speeds close to that of a parakeet for a few tenths of a second, but you can't call that long-distance cruising. Birds and planes are different: they travel best at speed.

Herring gull (*Larus argentatus*): $W = 9.4$ N, $S = 0.18$ m², $b = 1.4$ m.

Precisely because airplanes and birds are so different on this score, it is worth the trouble to find out how figures 6 and 7 would look if they were used to plot the performance of cars rather than birds. The equivalent of figure 6 is straightforward: you can find the necessary data neatly tabulated in any automotive yearbook. You know from experience that you pay dearly to drive faster than your neighbor. A tiny Fiat Panda, with a 20-kilowatt engine, has a top speed of 75 miles per hour, but if you want to drive 100 miles per hour you need at least 50 kilowatts. A speed of 200 miles per hour requires 300 kilowatts. A Porsche 959, with a 331-kilowatt engine, has top speed of 197 miles per hour. A Porsche 911 Turbo, with a 221-kilowatt engine, has a top speed of 160 miles per hour.

The rate at which the power required increases with increasing speed is illustrated best by another proportional diagram: figure 8. The engine power turns out to be proportional to the third power of the speed. Thus, if you want to drive twice as fast, your engine

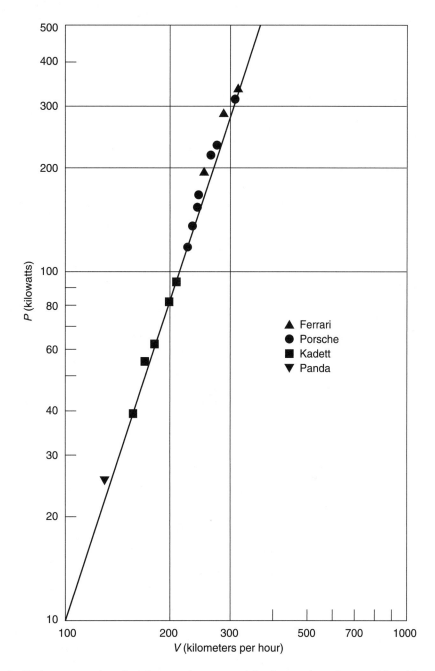

Figure 8 Engine power plotted against maximum speed for four makes of automobiles. The diagonal is steep because power increases as the third power of speed. To go twice as fast requires an eightfold increase in engine power.

has to be 8 times as powerful. A Porsche or a Ferrari can go 3 times as fast as a Fiat Panda, but its engine is 27 times as powerful. If you are determined to travel at high speed, you should take up flying. A sports plane with a large Porsche engine can easily reach 250 miles per hour.

Since the power P is equal to the drag D times the speed V, the data in figure 8 can be used to make a graph of the relation between D and V for cars, equivalent to figure 7. But we can do better. It is obvious that large birds and large vehicles consume more food or fuel than small ones. If you want to compare the energy needs of different modes of transportation, you must account for differences in size. A trucker would advise you to compute fuel consumption not per mile but per ton-mile. Drag is not the most relevant measure for the specific cost of locomotion; a better measure is the drag per unit weight, D/W. This is the quantity we shall focus on, bearing in mind that the payload of a vehicle is often only a fraction of its gross weight.

The quantity D/W registers energy consumption per meter traveled for each newton of gross weight. In thinking about the performance characteristics of different modes of transportation, this quantity is so important that it has a separate name: E, the specific energy consumption. In engineering notation,

$$E = D/W = P/WV. \tag{6}$$

The quantity E is "non-dimensional": if you stay with the metric system and measure both D and W in newtons, E is a pure number, which remains unchanged when you switch to a different set of units. Please note that P must be measured in watts and V in meters per second; otherwise one becomes hopelessly tangled in conversion factors.

Since the performance of Tucker's parakeet was always measured at the same weight (35 grams, or 0.35 newton), figure 7 remains unchanged if we convert the graph from D to E. The only change is the scale on the vertical axis. The scale for D is given on the left side of figure 7; that for E is given on the right. The best

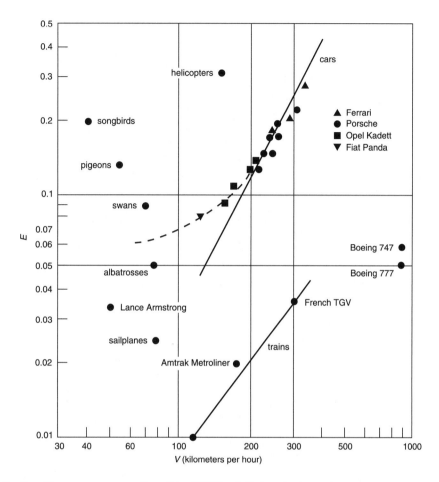

Figure 9 Specific energy consumption, $E (= P/WV)$, plotted against speed.

value for E that Tucker's parakeet could attain was 0.22, at a speed of almost 11 meters per second.

Cars can be dealt with in the same way. Consulting the automotive yearbook once more, and adding 200 kilograms to the empty weight for the driver, one passenger, fuel, and luggage, we can easily convert the data in figure 8 into data for E. The results are presented in figure 9, along with the corresponding numbers for passenger trains (commuter trains, the Amtrak Metroliner, and

France's TGV. Various birds and Boeings are not forgotten, either. Note that the drag of cars increases as the square of their speed. Most aerodynamic forces do; recall that the lift is also proportional to the square of the speed.

There is plenty of information in figure 9. For one thing, trains would appear to be far more economical than cars. However, when we account for the difference between useful load and gross weight, the economy of trains is not so clear-cut. A two-coach commuter train weighs 100 tons and seats roughly 100 passengers. At 70 kilograms a head, the people in a fully occupied train weigh about 7 tons—only about 7 percent of the gross weight. A Chevrolet with an empty weight of 2,000 pounds can carry four people, for a useful load of 600 pounds. This makes the payload equal to 23 percent of the gross weight $(600/(600 + 2,000) = 0.23)$—3 times the value for a train. At 75 miles per hour, a car's specific energy consumption, E, is about 0.08, and a train's is 0.011. Giving the train a markup for its excessive empty weight, we find that the corrected value of E is about 0.033, roughly half that of a car. These values reflect relative energy consumption per passenger-mile. Here are two easy numbers to remember: a train needs 1.6 megajoules per passenger mile (1 megajoule per passenger-kilometer), and a car needs twice that. But remember also that in terms of efficiency there is no difference between a fully loaded automobile and a half-empty train! Because the difference between cars and trains is not at all as significant as is often stated, the Netherlands State Railways—before privatization—used to sell discounted taxicab tickets to passengers. On the last leg of a train trip, a Dutch citizen could get a $5 ride in a brand-new Mercedes. A marriage between commuter trains and taxis sounds crazy but makes perfect sense. Sorry, but privatization is not necessarily always conducive to progress.

Songbirds can be found in the top left corner of figure 9, which shows that they are rather uneconomical, though not as bad as helicopters. Swans achieve a far better value of E, namely 0.09, which relative to cars is really quite good. Albatrosses and sailplanes are much more economical. Lance Armstrong ranks very well, too. (Ordinary bicyclists achieve $E = 0.01$, but at much lower speeds.) The most striking feature of figure 9 is that airliners do so well, though their speed is much higher than that of the fastest cars and trains. The Boeing 747 achieves $E = 0.06$ at 560 miles per hour. Its successor, the 777, does better yet: $E = 0.05$. (The elderly Boeing B-52 bomber, with its long wings, also achieves $E = 0.05$.) Airliners can retract their wheels once they are airborne; that option is not available to cars. Even with the most advanced streamlining, cars have awful aerodynamic properties, caused by the turbulence around the wheels and in the wheel wells. High-speed trains have the same drawback. Early airplanes could not retract their wheels, either. A Piper J3 Cub manages $E = 0.1$ at best, not nearly as efficient as a Boeing 777. Need I explain why all birds retract their legs after takeoff? Or why gulls use their webbed feet as air brakes?

A straight line drawn from songbirds to modern jetliners in figure 9 intersects the curve for cars in the vicinity of 200 kilometers per hour. Below that speed wheels perform better than wings, but at higher speeds wings have the advantage. Flying is the preferred mode of transportation when high speeds are desired. Most vehicles need additional power to achieve higher speeds, but a well-designed airplane can fly fast without using more fuel. This is why airships did not and will not survive evolutionary stresses. Their specific energy consumption is comparable to that of jetliners, but they fly less than 100 miles per hour. They produce only a fraction of the seat miles that airliners are capable of; therefore, their capital expenses per passenger mile are far too high. The Hindenburg disaster of 1937 ended an era; all later attempts were doomed to fail. A few years ago, the hangar for the giant German Cargo Lifter was converted to a tropical amusement park.

Nutrition and Combustion

So far we have not bothered to make a systematic distinction between fuel consumption and energy consumption, but now the time has come. Different foodstuffs and fuels supply different quantities of energy per kilogram consumed. There are also variations in the efficiencies of different energy-conversion processes. For example, a steam locomotive has an efficiency of only 5 percent, which means that of every 20 joules of energy available in anthracite or fuel oil only one joule is delivered as useful work to the driving wheels; the rest goes up in steam and smoke. The energy-conversion efficiency of human and animal metabolism is typically about 25 percent during strenuous work. A gasoline engine achieves a conversion efficiency of 25 percent at average speeds; nevertheless, three out of four joules of combustion heat are expelled by the radiator and the exhaust pipe, and only one out of four is available for propulsion. At 30 percent efficiency, diesel engines are a bit more economical. Today's best gas-fired electricity-generating plants manage 50 percent, much better than the 35 percent of 20 years ago. They achieve this by using the hot exhaust gases of the gas turbine to heat the boilers of steam turbines down the line. Modern high-bypass-ratio fanjet engines, such as those for the Boeing 777, achieve 50 percent too.

Consider an electricity-generating plant that uses natural gas with a combustion value of 36 megajoules per cubic meter. For such a big customer, the purchase price of natural gas is about $10 per 1,000 cubic feet, or 36 cents per cubic meter. This works out at about 1 cent per megajoule. Because half of the energy is wasted through the chimney or the cooling towers, the selling price of energy must be at least 2 cents per megajoule just to recover the purchase price of the gas. The user's monthly bill, however, is calculated in kilowatt-hours. A kilowatt-hour equals 3.6 megajoules; thus, in order for the utility company to break even, the selling price of electricity has to be more than 7 cents per kilowatt-hour. (Power companies calculate their off-peak rates along these lines. If investment costs can be recovered during peak periods, they are content if only fuel costs are covered at night.)

A megajoule is 0.28 kilowatt-hour. Reheating some leftover lasagna in a 700-watt microwave oven for 5 minutes uses about 0.2

Snow goose (*Anser caerulescens*).

megajoule of energy, at a cost of about 2 cents at the off-peak rate. The manual labor of cleaning up one's back yard requires about half a megajoule of nutrient energy every hour if one isn't working too strenuously—500,000 joules per 3,600 seconds, the equivalent of 140 watts. If you want to find out how many peanut butter sandwiches you will have to eat after an afternoon's work in your backyard, multiply those 500 kilojoules by the number of hours you have worked, then do the sums given below.

The nutritional value of peanut butter is stated on the label of the jar. Peanut butter is good for 180 "calories" per ounce. (These are actually kilocalories, so we're really talking about 180,000 calories per ounce.) With 4.2 joules per calorie and 28.4 grams per ounce, this works out to 2,700 kilojoules per 100 grams, or 27 megajoules per kilogram. Thus, a megajoule of peanut butter weighs about 37 grams, enough for three solid sandwiches. The price of this megajoule is approximately 20 cents, substantially more than the price of a megajoule of electricity. Electric trains use 1.6 megajoules per passenger-mile. (Just for fun, you can work out how much it would cost to run the train on peanut butter.)

Table 3 The heat of combustion or metabolic equivalent for various foodstuffs and fuels. The prices are based on a "snapshot" in 2008; large changes in fuel prices may occur over the years.

	Megajoules/ kilogram	Dollars/ kilogram	Dollars/ megajoule	Comments
Prime beef	4.0	32	8	
Beef	4.0	12	3	
Whole milk	2.8	1.20	0.43	600 cal/quart
Honey	14	4.70	0.33	
Sugar	15	1.30	0.09	100 cal/ounce
Cheese	15	10	0.67	
Bacon	29	10	0.34	
Corn flakes	15	7	0.47	100 cal/ounce
Peanut butter	27	7	0.26	180 cal/ounce
Butter	32	8	0.25	
Vegetable oil	36	3.5	0.10	240 cal/ounce
Kerosene	42	1.20	0.03	0.82 kg/liter
Diesel oil	42	1.00	0.024	0.85 kg/liter
Gasoline	42	0.90	0.021	0.75 kg/liter
Natural gas	45	0.80	0.018	0.8 kg/m^3

A carton of milk carries nutritional information, too. Milk supplies 600 kilocalories per quart (280 kilojoules of energy per 100 grams, or 10 megajoules per U.S. gallon). If milk costs $4 a gallon, a megajoule will set you back 40 cents. When you find out that the price of natural gas is about 2 cents per megajoule, you might almost be tempted to drink liquefied methane instead of milk. Steak is an outrage from this perspective. Its nutritional value is 4 megajoules per kilogram, roughly 2 megajoules per pound. At a price of $16 a pound for prime beef, that's about $8 per megajoule. Even the cheaper cuts of beef, at $6 a pound, cost as much as $3 per megajoule. Table 3 presents the nutritional values of several common digestibles and the heats of combustion of several fuels. In dollars per megajoule, natural gas and gasoline stand out as the best buys. At 65 cents per liter ($2.50 per gallon), gasoline in the United States costs about 2 cents per megajoule. (In Europe, gasoline costs about $6 per gallon, two-thirds of which goes to taxes.) Among digestibles, vegetable oil is cheapest by far. Salad oil is barely digestible; consumed in quantity it results in a bad heartburn. Peanut butter, though, is an excellent alternative for those who insist on getting a megajoule for their buck.

How to Lose Weight Quickly

How much food do birds need when they migrate south in the autumn? A mute swan, the largest of the European swans, cruises at about 20 meters per second (45 miles per hour). It weighs 10 kilograms (22 pounds), 2 kilograms of which constitute its flight muscles. In cruising flight, the power output of these flight muscles is about 100 watts per kilogram. During long-distance travel, therefore, the flight muscles of a swan supply approximately 200 watts of mechanical power. Because the conversion efficiency of nutrient energy is only about 25 percent, the swan needs 800 watts, or 800 joules of nutrient energy per second, during its flight. At a speed of 20 meters per second, this corresponds to an energy consumption of 40 joules per meter, or 64 kilojoules per mile. This energy is supplied by the spare fat on the bird's chest.

During a long flight, the pectoral muscles of birds metabolize fats directly. Human muscles, in contrast, burn sugars. In the human body, the liver must convert fats into sugars before the stored energy is of any use to the muscles. At the high metabolic rates typical of birds (remember, flying is plain hard work in most cases) this is not an attractive option. On top of that, the nutritional value of fat is twice that of sugar (table 3). Thus, it is much better to carry fat for fuel rather than sugar.

As always, there are exceptions. Hummingbirds run on honey and sugar water, and some marathon runners switch from sugar metabolism to fat metabolism in the course of a race (a process

White pelican (*Pelecanus erythrorhynchos*): $W = 60$ N, $S = 1$ m^2, $b = 2.80$ m.

that causes painful physiological changes). Most insects run on sugars, and so do chickens, partridges, and pheasants (which are not capable of long-distance flight). Migrating butterflies, however, store fat in their abdomens. The nutritional value of bird fat is about 38 megajoules per gram. Since a flying bird not only consumes fat, but also loses water and some body tissue, I prefer to use a somewhat lower number: 32 joules per gram. Since a swan requires 40 joules per meter, or 64 kilojoules per mile, it consumes 1.2 gram of fat per kilometer (2 grams per mile). After 12 hours of cruising at 45 miles per hour, the swan has traveled 540 miles and has lost more than a kilogram of body weight (1,080 grams, not counting the energy needed for its other body functions). Obviously a light snack will not fill its stomach at the end of such a working day. The same is true for homing pigeons at the end of a long-distance race: they are not only dead-tired, but famished as well. From this perspective, the importance of bird sanctuaries is easy to understand: migrating birds must eat voraciously before continuing their journey.

Gannet (*Sula bassana*): $W = 27$ N, $S = 0.25$ m^2, $b = 1.85$ m.

3 In Wind and Weather

The KLM flight that leaves London-Heathrow for Amsterdam at 6 p.m. each day is an airborne commuter train. The Boeing 737 is filled with regular customers: business people returning home from a day's work, familiar voices chattering to pass the time and teasing the long-suffering flight attendants. Many years ago, as I was dozing in my seat after a tedious meeting at the European Center for Medium-Range Weather Forecasts, the captain's voice on the public address system woke me up: "Ladies and gentlemen, this is the captain speaking. As you know, this trip usually takes about 45 minutes, but today is different. We have a 145-knot tailwind, which corresponds to 170 miles per hour. I have never before encountered such fierce winds. To compensate a little for the turbulence on this trip, we will arrive in Amsterdam 10 minutes early." (A knot is a nautical mile per hour, and a nautical mile equals 1.15 statute mile.) The 737 rode the center of the westerly jet stream, and the pilot played it to full advantage. The airspeed was 500 miles per hour, but because of the tailwind the ground-speed was 670 miles per hour—faster than the speed of sound.

When flying, you must always take the wind into account. You can't afford to be casual about it. Obviously it is convenient and economical when a tailwind helps you along, but you should be on guard when coping with headwinds. They set you back, and that cuts into your range. During World War II, American B-29 bombers on their way to Japan from Saipan and Tinian (islands just north of Guam) sometimes had to return prematurely because the headwinds they encountered were stronger than had been forecast. (There is a story of one bomber squadron actually flying backward on a particularly breezy day: fully opened throttles and

White stork (*Ciconia alba*): $W = 34$ N, $S = 0.5$ m², $b = 2$ m.

200-plus miles per hour were not sufficient to make headway against the storm.) Having consumed more than half the fuel in their tanks, they had no choice but to return. To lighten their planes and thus reduce fuel consumption, the crews dumped their bombs in the ocean. Once they had turned around, the headwinds became tailwinds, and so they got home in a hurry.

Wind influences airlines' timetables, too. The flight from Amsterdam to San Francisco takes about an hour longer than the return trip. Flying at 30,000 feet and up, jetliners encounter stiff westerly winds most of the time. Those mid-latitude winds are caused by the interaction of Earth's rotation and the temperature contrast between the equator and the poles. At cruising altitude, the average wind speed is about 30 miles per hour. Hence, during the 10-hour journey between Amsterdam and Los Angeles you lose 300 miles. Since long-distance jetliners travel about 550 miles per hour, this adds a little more than 30 minutes to the flight. Flying in the other direction, the wind works to your advantage. This explains the one-hour difference in the timetable.

The cooperation between meteorology and aviation benefits both parties. Airliners participate in the observation and measurement

Swallowtail (*Papilio machaon*): $W = 0.006$ N, $S = 0.003$ m^2, $b = 0.08$ m.

programs of worldwide meteorology, and the weather computers predict where the winds will be strongest. Every day, the most economical routes across the oceans are selected in the international conference calls of air traffic controllers. Westbound traffic often makes appreciable detours to avoid the strongest of the forecast headwinds. If the turbulence is not too severe, eastbound traffic is directed into the heart of the westerly jet stream. These adjustments reduce both travel time and fuel consumption. In the crowded skies above Europe and the United States, however, airline traffic is so congested that everyone must stick to the appointed airways. Only coast-to-coast nonstop flights are assigned the routes with the best winds.

The slower you fly, the more the wind will affect you. The large propeller-driven airliners of the 1950s, the Douglas DC-7 and the Lockheed Constellation, cruised at about 300 miles per hour and had to stop for fuel at Gander, Newfoundland, if they ran into unexpected headwinds over the Atlantic. Today, with cruising speeds twice as high, stopping at Gander is a thing of the past.

What is true for airplanes is also true for birds. Because birds are much smaller than planes, they fly more slowly, and this makes them vulnerable to adverse weather. The average wind speed on Earth is between 5 and 10 meters per second (between 11 and 22 miles per hour). If they want to return home as a storm approaches, birds must be able to fly about 10 meters per second, and their

Table 4 The Beaufort scale. The cruising speeds of various insects, birds, and airplanes are given for comparison.

Beaufort no.	Airspeed	Windspeed (m/s)	Cruising speed of
1	Light air	0.5–1	Butterflies, damselflies
2	Light breeze	2–3	Gnats, flies, dragonflies
3	Gentle breeze	4–5	Human-powered airplanes
4	Moderate breeze	6–8	Bees, wasps, beetles
5	Fresh breeze	9–11	Sparrows, starlings, swallows
6	Strong breeze	11–13	Crows, gulls, falcons
7	Near gale	14–17	Plovers, knots, godwits
8	Gale	18–21	Swans, ducks, geese
9	Strong gale	21–2	Sailplanes
10	Storm	25–28	Home-built aircraft
11	Violent storm	29–32	Diving hawks
12	Hurricane	> 32	Diving falcons

wing loading must range between 10 and 100 newtons per square meter. (See figure 2.) Most birds weigh between 10 grams and 10 kilograms, which puts their wing loading in the desired range. The vertical line in the center of figure 2 was drawn for a good reason: birds that fly faster have some speed to spare when the wind increases; the rest must watch out.

Still, one can easily imagine circumstances in which the wind blows so hard that all birds must seek shelter. This happens sooner for small birds than for large ones. (See table 4.) Storm petrels were so named because they are the first ocean birds to seek refuge ashore as a storm moves in. Their arrival above land is an early warning signal. Small birds have low wing loadings and hence low airspeeds. As a consequence, they must seek shelter sooner than their larger relatives. The smallest birds, including goldcrests

Griffon vulture (*Gyps fulvus*): $W = 70$ N, $S = 1$ m^2, $b = 2.60$ m.

and kinglets, cannot survive the open plains or the ocean shores. Their habitat is the forest, where trees and shrubs protect them from high winds.

Insects are so small that their lives are dominated by the wind, and the very smallest must wait for the wind to die down at sunset before taking to the air. Gnats start their dance in your back yard early in the evening, else they would be carried off by gusts. The insects in figure 2 can be divided into two categories: those with high and those with low wing loadings. Beetles, flies, bees, and wasps belong to the first category, butterflies and dragonflies to the second. Honeybees can return to their hive after harvesting nectar and pollen in a faraway rapeseed or alfalfa field. Butterflies, however, must accept that they may be blown away by the wind, and dragonflies cannot venture far from their hideaways in windy weather. But accidents do happen occasionally. When blown off their track by an easterly storm, grasshoppers from the Sahara sometimes turn up in England. More often they drown in the Atlantic Ocean, exhausted.

The difference between a maritime climate and a continental one is considerable. Ocean birds live in a windy environment, which explains why most of them are fairly large. Large birds have higher wing loadings and airspeeds than small ones, and this is a major advantage at sea. Because ocean birds must cover great distances, it is essential that their energy consumption per mile be low. Their long, slender wings achieve just this. The narrow wings of seagulls, terns, and albatrosses are quite different from the broad wings of vultures, condors, and eagles. Those large birds of prey do not travel long distances; they soar in circles, taking advantage of the ascending motion in thermals. In thermal soaring the energy consumption per unit distance does not matter. The wings of soaring birds of prey minimize the energy consumption per unit time (that is, the rate at which energy must be extracted from the air). This goal is achieved by flying slowly; hence the low wing loadings. The enormous wings of the golden eagle make perfect sense.

The Art of Soaring

Flying is an arduous way of life. This is why several species of birds have discovered how to stay airborne without flapping their wings. The trick is to find upward air movements of sufficient strength. Under normal circumstances a bird glides down when it doesn't flap its wings; then, as it loses altitude, gravity supplies the energy needed to maintain airspeed. But when the rate at which updrafts lift a bird is equal to its rate of descent, the bird can stay up indefinitely. And if the upward air motion is stronger than the bird's sinking speed, it can gain altitude if it wishes. Staying aloft without having to work for it is the art of soaring.

There are several ways to soar. One is practiced by herring gulls as they follow a ferry or a cruise ship. They fly on the windward side of the ship, where the wind escapes upward as it strikes the superstructure. The gulls need only adjust their wing area. When

Figure 10 Slope soaring alongside a ferry and along a dune ridge.

the wind increases, they fold their wings a little. If they lose too much altitude, they spread their wings again. If they start going too fast, they simply extend their feet a bit more, and that slows them down. Webbed feet are perfect air brakes. (Every glider pilot knows how important it is to increase drag briefly when flying too fast or too high.) The flight of seagulls alongside a ferry is called "slope soaring." It can also be practiced along chains of dunes and mountain ridges. But there must be sufficient wind, because the upward velocity along the slope is proportional to the wind speed. Furthermore, the wind should blow across the ridge or it will not be diverted upward (figure 10).

In favorable circumstances, gulls and terns can soar back and forth for very long periods without ever flapping their wings. They are getting a free ride, and what a ride it is! Kestrels and harriers (sparrow hawks and marsh hawks to some) do the same inland, along the slopes of hills and levees. Hang gliders and sailplanes also take advantage of slope winds. Since hang gliders have a sinking speed of more than 2 meters per second, they need a stiff breeze before they can venture flying along the shore.

Slope soaring is perfect for covering large distances. Glider pilots achieve their distance records on days when high winds are blowing across long mountain ridges. The narrow northeast-southwest folds of the Appalachian Mountains stretch from Elmira, New York, to Chattanooga, Tennessee. When a storm hits that 600-mile stretch, stirring up blustery northwesterly winds behind its cold front, complete with deep-blue skies and towering cumulus clouds, glider pilots in Elmira ("soaring capital of the world") are eager to get airborne.

The second method of soaring capitalizes on thermals. This is not without its drawbacks. For one thing, thermals do not occur everywhere, and you are not going to get anywhere unless you can first find a thermal in which you can gain altitude. You do that by circling in the hot, rising column of air. Once you have climbed the thermal for a while, you can start your journey by gliding in the general direction of your destination, hoping to find the next thermal before losing too much altitude. The flight toward your goal will be punctuated by such episodes in the winding staircases of rising air. This kind of soaring is possible only during the day, because thermal motion occurs only after the sun has begun to heat Earth's surface. You can observe this by watching buzzards (buteos, to some) and other soaring birds of prey. As the morning progresses, these birds test the strength of the convection currents. They take wing, searching for ascending air. If they fail to maintain altitude, they return to their tree or rock ledge and wait for the surface to warm up a bit more. The flight muscles of soaring birds of prey cannot sustain flapping flight for more than a few minutes.

As hot air rises, it cools—about 1° Celsius per 100 meters. At a certain altitude, the water vapor in the ascending air begins to condense; the cumulus clouds created that way are a sure sign of upward motion. For this reason, glider pilots join the gulls and hawks that are circling below these clouds when they want to gain altitude. The rates of descent of soaring birds and gliders are comparable (around 1 meter per second), but the airspeeds of birds are lower. Birds can keep up with their fiberglass companions only by flying in tighter circles. If a bird and a glider pilot both are in a playful mood, or have a researcher's attitude (which amounts to the same thing), they may start flying competitively to see who dares to turn the tighter circles, who can fly slower without losing control or stalling (literally dropping out of the race in that case), who has the smarter tactics for discerning the next thermal and reaching it with minimum altitude loss, and so on. Buzzards appear to take these games very seriously. So do scientists: the soaring habits of griffon vultures in Central Africa were investigated by a biologist in a glider with auxiliary engine.

Long-Distance Migration

Birds cover enormous distances on their annual migrations. Unless they forecast the weather well, they risk running into serious trouble. Since their cruising speeds are relatively low, they will consume too much fuel if they run into headwinds. When the wind shifts direction, they risk being blown off course, with fatal consequences if they should end up over the open ocean. Careful preparations are necessary before they start on their journey. This is especially true for small birds, such as the North American passerines (chimney swifts, bank and cliff swallows, purple martins, blackpoll warblers, redstarts, and the like) that cross the Gulf of Mexico on their way south. Their performance should not be underestimated: 500 miles nonstop is quite a feat. But the real heroes of Gulf migration are the monarch butterfly and the ruby-throated hummingbird. Both of them are known to consume fat on their way across the Gulf, and both convert sugar into fat in preparation for the trip.

When I drafted the first edition of this book, I assumed, on the authority of ornithologists, that all birds that choose to cross the Sahara desert directly, instead of making detours over Israel or Gibraltar, do so nonstop. Massive amounts of recent radar data prove that I was mistaken. Little songbirds typically fly at night, and hide from the daytime heat in shadowy spots along the way, in order to minimize evaporative water loss. Since they cannot fatten up along the way, they must store enough fat for the entire 1,000-mile Sahara crossing.

The little European passerines that migrate to Africa for the winter, and have to cross both the Mediterranean Sea and the Sahara, include various kinds of warblers, wagtails, pipits, chiffchaffs,

Common buzzard (*Buteo buteo*): $W = 10$ N, $S = 0.27$ m^2, $b = 1.35$ m.

flycatchers, redstarts, and nightingales. With a migration speed of only 10 meters per second (22 miles per hour), they wait for strong tailwinds before they start their crossing. Little birds consume relatively large amounts of energy. For passerines, just as for Tucker's parakeet, the specific energy consumption ($E = D/W$) is approximately 0.25. At cruising speed, therefore, their aerodynamic drag is about one-fourth their weight, which is not particularly economical. Because all these birds cross the Sahara without a chance to forage, their cruising range must be at least 1,000 miles. They manage this by storing so much fat on their chests that they can barely fly. A 20-gram wagtail, with a normal fuel reserve of 5 grams, starts its journey across the Sahara with an additional 10 grams of fat. Its takeoff weight is 30 grams (a little more than an ounce)—twice its zero-fuel weight. Half of its takeoff weight is fuel, much as for a long-distance airliner. Because songbirds must fatten themselves

Flamingo (*Phoenicopterus ruber*).

in preparation for their flight across the Mediterranean, and because they have trouble taking off with so much excess weight, they are easy prey for Maltese bird catchers. (A roasted nightingale tastes much better when its meat is wrapped in fatty tissue.)

Let's take the average weight of a wagtail on a long-distance flight to be 24 grams, and assume that its drag is one-fourth of its weight. Its average drag then is 6 grams, or 0.06 newton. As was explained in chapter 2, a newton is a joule per meter. If we can compute how many joules of mechanical energy are supplied by metabolizing 15 grams of bird fat, we can calculate the wagtail's maximum range. Bird fat supplies 32 kilojoules per gram (chapter 2); hence, 15 grams supply 480 kilojoules. However, since the bird's metabolic efficiency is only about 25 percent, the net supply of mechanical energy is no more than 120 kilojoules. This is used up at a rate of 0.06 joule per meter, or 0.06 kilojoule per kilometer. After 2,000 kilometers (1,250 miles), all the fat is gone.

With fuel reserves for only 250 miles on a 1,000-mile trip, there is not much to spare. Airliners don't cut it that close! Birds cannot afford any miscalculations in their weather forecasts. They are wise enough to wait for the wind to blow in the right direction. Circling over their feeding grounds each morning, they check the meteorological conditions. The starting signal for the great journey is given only when conditions are favorable. Even with a strong tailwind, the flight across the Sahara takes at least two days. Wagtails happen to fly during daytime and rest at night, unlike most other passerines. If the weather suddenly deteriorates during the trip, many of the migrating birds may die. Similar risks are taken in crossing the Gulf of Mexico. Few passerines survive when they are blown far out into the Atlantic by unexpected westerly gales. For many centuries sailors have told stories of songbirds escaping their fate by dropping on the deck of a ship, famished and exhausted, then recuperating quickly on bread crumbs, scraps of bacon, and some tender loving care.

Some bird species do make nonstop flights across the Sahara. Even one species of gull does. Swiss ornithologists brought their radar equipment to the inland deserts of Mauritania to observe the spring migration of lesser black-backed gulls (*Larus fuscus*). These are big birds, comparable in size to herring gulls. Their weight is 7.2 newtons, their wingspan 1.34 meter, and their wing area 0.1934 square meter. This gives them a calculated cruising speed of 10 meters per second (22 miles per hour). The airspeed observed by the Swiss radar crew is only a little faster: 11 meters per second (24 miles per hour). These gulls start from the Atlantic shores near Dakar, Senegal, and proceed to climb to 3,500 meters, where 22-mph tailwinds make their ground speed twice their airspeed. With this much tailwind they need not fly at top speed, so they don't. But it does mean they can cross 3,000 kilometers (2,000 miles) of desert in just 48 hours. This is what they do, nonstop from Dakar to the shores of the Mediterranean Sea near Algiers. They do not have to fatten up much before departure, because they get 1,000 miles free. Less than 100 grams of fat would do. They also don't

Osprey (*Pandion haliaetus*): $W = 15$ N, $S = 0.3$ m^2, $b = 1.60$ m.

have to worry about overheating and water loss, because it is cool enough at altitude. Still, their performance is a marvel. These ocean birds would be at a loss if they had to land in the desert for a stopover. (This daring strategy is impossible during autumn migration, because it is hot over the desert at the end of the summer, and the northeasterly trade winds can be exploited only when flying low. The gulls therefore follow a route closer to the Atlantic coast on their return to Africa in the autumn, and make a number of stopovers then.)

Some migrating ocean birds cover distances much greater than the 500 miles across the Gulf of Mexico or the 1,000 miles across the Sahara. Several species of plovers, godwits, and sandpipers make nonstop trips from Cape Cod to Trinidad, a distance of 2,500 miles. Since this kind of flying requires much more advanced aerodynamics, the bodies of ocean birds are streamlined and their wings slender. As a result, the ratio between drag and lift decreases substantially: for ocean birds it is about 0.07, instead of the 0.25 typical of songbirds. Wilson's phalarope, a little shore bird that migrates more than 5,000 miles along the Pacific coast of the Americas, prepares for its annual journey by filling its belly with brine shrimp in Lake Mono, California. Like migrating passerines, it fattens itself until it can barely fly. Because the ratio $E = D/W$

remains roughly the same, its aerodynamic drag increases by 50 percent when its weight grows to 50 percent above normal. Carrying that much excess weight, yet using the same wings, the phalarope must fly more than 20 percent faster. The power required equals drag times speed ($P = DV$; see chapter 2); it increases to almost twice the normal value ($1.5 \times 1.225 = 1.84$). If a phalarope continued stuffing itself, it wouldn't be able to take off at all. Once airborne, however, with such an economical value of E, it flies more than 3,000 miles nonstop. Even so, it cannot make the journey from California to Chile without stopping for fuel along the way. From time to time it has to forage for shrimp and other seafood along the beach.

Each species has its own strategy. The sandpipers that breed along the coast of Greenland and pass the beaches of northwestern Europe each August, en route to destinations further south, appear to have learned that westerly winds are generally stronger at high altitudes. On their 1,200-mile journey to Scotland, they have been observed flying as high as 7 kilometers (23,000 feet)—almost above the weather. For the same reason they skim the waves on their journey back to Greenland, riding on the easterly surface winds blowing north of mid-latitude storms.

Plovers, Knots, and Godwits

Let's take a closer look at the performance of three species of wading birds. Red knots (*Calidris canutus*) start their spring migration from the shores of Mauretania to northern Siberia against the subtropical trade winds along the African west coast. How do they manage that, all fattened up? No bird flies into a headwind when it doesn't have to. Ringed plovers (*Charadius hiaticula*) have to cross the Greenland ice cap on their way to the Dutch Wadden Sea. How do they climb 10,000 feet to clear the snow-covered highlands there? In autumn, bar-tailed godwits (*Limosa lapponica*) make nonstop flights from Alaska to New Zealand, a distance of 11,000 kilometers (7,000 miles)! When I drafted the first edition of

Monarch butterfly (*Danaus plexippus*).

this book, I did not dare to conceive of such feats. I was convinced that nonstop flights of 5,000 kilometers were the very longest one could expect. Available data on flight muscle efficiency, aerodynamic drag, and flight speed would have ruled out such daydreams. But we realize fully now that the flight performance of wading birds is about twice as good as professionals have thought for more than 30 years.

The red knot is a species that has been thoroughly investigated. Theunis Piersma of the Netherlands Institute for Sea Research will be remembered forever because a subspecies of knot was named after him. He and his Swedish colleagues Åke Lindström and Anders Kvist studied the physiological changes in knots in preparation for and during migration. They discovered that knots "burn their engine" during their journey, decreasing their muscle mass as they lose weight during the trip. They also found that knots engage in "body building without power training." Their flight muscles double in weight before they start their trip. Other physiological change are drastic, too. Knots shrink their digestive organs to a minimum in the days before takeoff, and their hearts and pectoral muscles shrink as they proceed on their journey. In fact, Piersma says, forgetting about their bones and feathers, "only their brains don't change during the trip." Piersma's Swedish colleagues supplemented these investigations with wind-tunnel work. Knots seem fairly comfortable in a wind tunnel, and will fly for hours on end as long as an experimenter remains at the window to keep them company. Still, they fooled the Lund University staff with rather sloppy metabolism. On migration, they have to treat every gram of fuel with care, but in a wind tunnel their apparent muscular efficiency is hopelessly low.

In any case, suppose you are a knot foraging on the wetlands of Mauretania, a little north of Dakar on the west coast of Africa, and preparing to take off for the 5,000-kilometer journey to the wetlands of Holland. Your weight has grown to 2 newtons (200 grams) and your flight muscles have grown to 45 grams. You have 60 grams of fat wrapped around your chest and belly. With a wing area of 0.0286 square meter, your most economical speed (according to equation 2 in chapter 1) computes as 14 meters per second. But the trade winds blow at least 10 meters per second, and right from the direction you have to head for. Flying against such a headwind clearly is no option. So you have to climb to 10,000 feet, where counter-tradewinds prevail, which will give you an appreciable tailwind. Biological evolution has programmed you to strengthen your muscles for this exact reason. Forty-five grams of flight muscle provide 4.5 watts of mechanical power. At a speed of 14 meters per second, you need only 2 watts for level flight; you can use the rest for climbing. Lifting a weight of 2 newtons at a rate of 1 meter per second requires 2 watts. That is what you choose; you don't want to go to the limit.

You are now at 10,000 feet above Morocco's coastline, with a fair tailwind. Your muscles are big and strong, and you decide to take advantage of their good condition. So you speed up to 20 meters per second (45 miles per hour), a speed at which your aerodynamic performance is somewhat worse than before. Your drag increases to one-twelfth of your weight (that is, 0.17 newton), so you now require 3.4 watts of muscle power. There is some power to spare;

Ruddy turnstone (*Arenaria interpres*).

before too long you will begin to consume both your heart and your muscles bit by bit. By the time you arrive at Holland's shores, you have slimmed down to 120 grams and you have only 20 grams of flight muscle left. You still race at 20 meters per second, which at this weight is quite at lot faster than the economical cruising speed. However, your body has slimmed down considerably, so its drag has become less. This way, your total drag is still one-twelfth of your body weight (0.1 newton). Therefore, the power required is 2 watts. You are evidently going to the limit, because your muscles produce only 2 watts at the end of the trip. However, the end is in sight, and you can drop out of the sky completely exhausted. Within a few weeks, you are off again for the next 5,000-kilometer stage of your journey, heading for the breeding grounds on the Taymyr Peninsula of northeastern Siberia.

But did you have enough fuel on board to fly 5,000 kilometers nonstop? You have lost 80 grams of weight, mostly fat but also some muscle. At an average metabolic value of 32 kilojoules per gram, you have consumed about 2,500 kilojoules of energy. If your energy-conversion efficiency is 25 percent, that corresponds to 625 kilojoules mechanical. How much work did you need to perform to overcome aerodynamic drag? Your average weight during the trip was 1.6 newton, and your drag was one-twelfth of that (0.133 newton). Force times distance equals work. Thus, over 5,000 kilometers you used up 667 kilojoules, somewhat more than you had available. Fortunately, tailwinds helped to overcome this discrepancy. In case of bad luck, you would have made a fuel stop halfway through.

In this story I have used approximate performance numbers based on experimental results obtained in recent years. The aerodynamic drag of wading birds is not one-eighth of their weight, but one-twelfth at the high speeds typical of their migration. And the metabolic efficiency of their muscles is not in the neighborhood of 10 percent, as a number of wind-tunnel experiments seem to suggest. Human long-distance athletes manage 25 percent; why should the performance of traveling birds be inferior?

These numbers find support in the entirely unexpected behavior of ringed plovers, which breed on the tundra of Baffin Island, on Canada's east coast. In early autumn, they return to their winter home on the Dutch Wadden Sea, and on their way they have to clear the high dome of the Greenland ice cap. Naturally, they start their trip with strong westerly winds. When these winds hit the high cliffs of Greenland's west coast, they cause appreciable slope winds. What do the plovers do? They stop flapping and start soaring! If their drag is one-fourteenth of their weight at a flight speed of 15 meters per second, vertical wind speeds of more than 1 meter per second will lift them over the cliffs without their having to spend energy. In other words, these plovers now act as if they were gulls! This behavior would not occur if their flight performance was as poor as it was long thought to be.

This brings me to the bar-tailed godwits that cross the entire Pacific Ocean without refueling. As waders go, these are large birds, with a takeoff weight of 500 grams, a wingspan of 0.73 meter, and

Franklin's gull (*Larus pipixcan*): $W = 2.5$ N, $S = 0.08$ m^2, $b = 0.95$ m.

a wing area of 0.0573 square meter. With these numbers, the economical cruising speed computes as 15 meters per second, just like that of other waders. Their flight muscle mass at takeoff is 70 grams, providing 7 watts of power. How much power do they need? If their lift-to-drag ratio at takeoff is 12 (I chose a fairly conservative number because they are so fat), their drag is 0.42 newton and the power required is a little more than 6 watts. Godwits are cutting it close; they have virtually no power left for climbing. Fortunately, they don't need it, because they need not pass mountain ridges and are smart enough to wait for the fierce northerly winds that prevail right after the passage of the first big autumn storm. Off they go. By the time they arrive over the coasts of New Zealand and the Northern Territory of Australia, their weight is down to 220 grams. They have consumed 220 grams of fat, 40 grams of flight muscle, and 20 grams of other body tissues. The metabolic energy consumed is about 9,000 kilojoules. With a conversion efficiency of 25 percent, this becomes about 2,300 kilojoules mechanical. Their average weight during the trip is 3.6 newtons, and with a lift-to-drag ratio of 14 their average drag is 0.26 newton. With these numbers, their flight range computes as 8,800 kilometers (5,500 miles)—2,200 kilometers short of the 11,000 kilometers needed.

There is some explaining to do here. How much do these godwits gain in the tailwinds of the first day after takeoff? Winds of 70 kilometers per hour are not uncommon in the Northern Pacific, giving the birds a 1,700-kilometer advantage in the first 24 hours. Not enough, though. But we can fiddle with the sums. I picked an optimistic number for the aerodynamic drag, but I could easily be less conservative about the meteorology these birds have mastered. If they manage to ride the northeasterly trade winds north of the equator, and climb high enough to profit from the counter-trades in the southern hemisphere, they might gain another 1,000 kilometers. Also, the energy-conversion efficiency of their flight muscles might be better than I assumed. Some human athletes reach 30 percent; what if godwits manage that, too? Available mechanical energy then computes as 2,700 kilojoules, and the flight range as 10,300 kilometers. With 1,700 kilometers of wind assistance there are even 1,000 kilometers to spare on arrival. The tank isn't emptied to the last drop, so to say.

The point of these exercises, however, is not to obtain precise numbers. That is impossible anyway if one thinks of all the factors that may affect a 7,000-mile flight. Instead, I am doing these sums in order to confirm that recent data on the migration of wading birds require a drastic revision of conventional wisdom on aerodynamic performance and metabolic efficiency.

Taking Off and Landing

Whenever they can, birds and airplanes take off and land into the wind. They need speed in order to become airborne. It is the speed with respect to the air that matters, not the speed with respect to the ground. Pilots speak of airspeed versus groundspeed. When the wind blows, groundspeed and airspeed are not the same. Franklin's gull and the similar European black-headed gull have

Rock dove (domestic pigeon, *Columba livia*): $W = 2.8$ N, $S = 0.07$ m², $b = 0.78$ m.

an airspeed of nearly 10 meters per second. That is 22 miles per hour, or force 5 on the Beaufort scale. With a headwind of 10 meters per second, a gull makes no headway at all. This is a nuisance if one has to get somewhere, but it becomes an advantage during takeoff and landing. A gull perched on a warehouse roof or a harbor bollard in a stiff breeze has only to spread its wings to obtain the lift required. A little hop into the air, a few casual wingbeats, and away it flies, with no effort at all. Taking off in calm weather is not so easy. The gull can either dive off its perch or take off vertically with rapid beating of its wings. The second option is hard work, requiring 4 times as much power as ordinary flight. This is why most birds prefer to take off from a tree, a telephone pole, a gutter, or some other elevated object. Starting with a brief dive, the bird gains the necessary airspeed by letting gravity do the work.

Landing works in much the same way. If circumstances allow, a bird lands into the wind, because that minimizes groundspeed. This is why a bird can land on a fence quite casually, as though it doesn't require great precision and exacting coordination. A pigeon lands on a roof ledge or a windowsill by deliberately approaching the landing spot from below and sailing gracefully upward in the last few meters, losing speed on the way up until its flight speed drops to zero at the chosen spot.

If a bird has to cope with a tailwind when taking off, it is in trouble. Because its groundspeed is now higher than its airspeed, it must run like crazy before its airspeed is high enough for takeoff. When the wind comes from behind, a bird must make an extremely long takeoff or landing run.

Airliners also have to worry about the wind, though not as much as birds, because their speeds at takeoff and touchdown—about 200 and 150 miles per hour, respectively—are much faster than typical wind speeds. Nevertheless, no airplane can afford to take off or land with the wind at its back.

The runways of major airports are usually about 2 miles long. Is that long enough for the takeoff run of a wide-body jet? An airplane can lift itself off the ground only after achieving sufficient airspeed. It must be accelerated before it can fly, and to do the necessary calculations we need to know the plane's acceleration when all its engines are running at takeoff power. The jet engines of a modern airliner deliver a total thrust equal to roughly one-fourth of the takeoff weight. Not all of this thrust can be used for acceleration, however; we have to make an allowance for the average aerodynamic resistance during the takeoff run. Therefore, we estimate the net thrust to be 20 percent of the takeoff weight. With T standing for thrust and D for drag, we have

$$T - D = 0.2W. \tag{7}$$

Because force times distance equals work, the product of the net thrust and the length R of the takeoff run is the work performed by the engines. This work is converted without any loss into the kinetic energy of the airplane. You may remember from your high school science class that the energy of motion, or kinetic energy, is $\frac{1}{2}mV^2$, where m is the moving object's mass and V is its speed. The last thing we need here is the relation between mass and weight. The weight W is the force exerted on an object of mass m by the pull of gravity. If we call the acceleration of gravity g, we can write

$$W = mg. \tag{8}$$

With the aid of equation 8, we can write the kinetic energy as

$$K = \tfrac{1}{2}mV^2 = \tfrac{1}{2}(W/g)V^2. \tag{9}$$

The energy supplied by the engines equals the net thrust times the length R of the takeoff run. Using equations 7 and 9, we obtain

$$\tfrac{1}{2}(W/g)V^2 = 0.2WR. \tag{10}$$

This can be simplified. When we divide both sides of equation 10 by W and multiply both sides by $2g$, we find

$$V^2 = 0.4gR. \tag{11}$$

Computing the length of the takeoff run now becomes easy. With $g = 10$ meters per second squared and $V = 84$ meters per second (190 miles per hour), we obtain $R = 1,764$ meters (almost 5,800 feet; a little more than a mile). However, in order to provide an adequate margin of safety, a runway must be roughly twice as long as the takeoff run. Should one of the engines fail during takeoff, an airplane should still be able to stop at the far end of the runway. Therefore, an airplane requiring a 1-mile takeoff run needs a 2-mile runway. This, incidentally, is a good example of the design philosophy used in aviation safety calculations. Fair margins, neither too large nor too small, are incorporated to allow for adversities that rarely arise.

Sandhill crane (*Grus canadensis*): $W = 45$ N, $S = 0.5$ m^2, $b = 2$ m.

When we start adding realistic details, calculations such as those described above become rather involved. The maximum speed at which a takeoff run can be safely aborted depends on the gross weight of the airplane and several other factors, such as the air temperature (in hot air, jet engines deliver less thrust). The co-pilot consults the airplane's manual to find the precise number, and when the "decision speed" is reached he or she lets the captain know. Beyond this point it is impossible to brake to a full stop. The plane is now committed to taking off, even if one of the engines fails. A few seconds after reaching decision speed, the pilot pulls the nose up, and a few seconds after that the airplane leaves the ground. The risk of engine failure was one of the reasons why intercontinental airliners had to have at least three engines (witness the awkward position of the tail engine on the almost forgotten Lockheed Tri-Star and Douglas DC10). This requirement was gradually dropped after 1990. A Boeing 777 with only one of its two engines operating properly can still take off safely, though it cannot climb very fast anymore and it is not allowed to continue its journey. It is a tribute to the tremendous reliability of modern jet engines that all long-distance airliners, including many types with two engines, are allowed to cross the oceans nowadays.

For more heavily loaded airplanes, the margins become narrower. A 747-400 with a takeoff weight of 380 tons must accelerate to 210 miles per hour before it can become airborne—20 miles per hour more than the speed that was mentioned a moment ago. With $V = 93$ meters per second (210 miles per hour) instead of 84 meters per second (190 miles per hour), we compute $R = 2{,}160$ meters (almost 7,100 feet). At that point, a 10,000-foot runway has only 2,900 feet left. No wonder the decision speed is much lower than the takeoff speed in this case: only 180 miles per hour. At this point, about 5,000 feet of runway have disappeared under the wheels.

It is easy to compute how wind speed affects a takeoff run. If there is a 30-mile headwind, the airspeed of 210 miles per hour needed by a fully loaded 747 for liftoff is reached at a groundspeed of 180 miles per hour. The takeoff run is then reduced from 7,100

White stork (*Ciconia alba*): $W = 34$ N, $S = 0.5$ m^2, $b = 2$ m.

to 5,200 feet. While you are checking this, using equation 11, please take a moment to consider the opposite situation. With a 30-mph tailwind during takeoff, the groundspeed would have to be $210 + 30 = 240$ miles per hour before the airspeed would be high enough for liftoff, requiring a ground run of 9,300 feet. This would leave precious little runway to spare. (Don't worry; no pilot would ever try this.)

It is also advantageous to land into the wind, of course. In a 30-mph headwind, a Fokker F100, with a landing speed of 120 miles per hour, has a groundspeed of only 90 mph when it touches down, which shortens the landing run considerably. For this reason, air traffic in the vicinity of an airport is always arranged in such a way that all aircraft take off and land into the wind. Diverting traffic to the runway causing the least noise pollution is possi-

ble only when there is little wind. As airliners preparing to land at
London Heathrow skim their rooftops, the citizens of the London
suburb of Hounslow are forcefully reminded of the prevailing
westerly winds.

Approach Procedures

Rigid procedures must be observed in order to achieve orderly air
traffic around airports. The usual preparation for the landing se-
quence begins with a descent to 5,000 feet. The pilot then receives
radar vectors to a point about 12 miles "downwind" (that is, paral-
lel to the intended runway but in the opposite direction). Then the
plane makes a 180° "procedure turn" to align itself with the runway. After this last turn, flaps and landing gear are extended and
airspeed is reduced to about 150 miles per hour, or 2.5 miles per
minute. Since there is a distance of about 12 miles to cover before
touchdown, this segment of the flight, which is called the "final
approach," takes about 5 minutes. Most passengers wouldn't mind
having this part sped up a little, but for pilots the last few minutes
before touchdown are very busy. A hurried approach, with a steep
turn just before touchdown, would cause great stress in the cockpit. Stunts like that are best left to fighter pilots and barnstormers.

It is not only pilots who need to practice approach procedures
until they become routine. Birds must do the same thing. In my
college years I used to go to summer camp on an island off the
Dutch coast. There was plenty of time for bird watching, and once
I saw a mature herring gull teaching his fledgling son step by step
what is involved in making a smooth landing. First, choose your
landing site and watch the waves for the wind direction. Next,
monitor the crosswinds that drive you off course and fly downwind for a while before making a turn into the wind. Now real
skills are needed. Stop flapping your wings, start your descent at a
speed that allows you to cope with wind gusts, monitor your descent with reference to your landing site, extend your legs a little
if you are not descending fast enough, reduce your speed during

Cape pigeon (*Daption capensis*): $W = 4.3$ N, $S = 0.077$ m^2, $b = 0.88$ m.

the last few seconds of the approach by leaning back and keeping your nose high, decelerate all the way by spreading your wings and tail fully, lower your landing gear, lift your wings above your shoulders, touch down, and fold your wings.

The juvenile gull, easy to spot because of its mottled gray and brown feathers, did its best to follow father's example, but with mixed success. On that afternoon it already knew the difference between downwind and final approach; however, it couldn't manage to fly a smooth approach after the final turn. Sometimes it would come in too high, sometimes too low. When its speed was too high, the youngster tried to correct by leaning back rather than extending its legs. What happened then? Lift increased, and the bird would soar upward until it realized that it was well above its glide path. A steep dive followed: wrong again. Diving generates excessive speed. If you try to fix that by pulling up again, you will find yourself too high for a second time. All very reminiscent of a student pilot early in flight training. This rather undignified performance often culminated in pure embarrassment. Despite its efforts, the young gull did not manage to make a single smooth touchdown. Sometimes it would land too fast and would trip over its feet, performing an accidental somersault. Then it would try to imitate its father's gentle "flare-out." (By leaning back and fully extending its wings and legs just before touching down, a bird loses speed without gaining altitude. At the moment its lift and its speed drop to zero, its feet should be only an inch or so above the

Laughing gull (*Larus atricilla*): $W = 3.3$ N, $S = 0.01$ m^2, $b = 1$ m.

beach. Now the bird simply falls, but since it has performed this trick so well it touches the sand with a barely perceptible impact.) But the young gull continued to overreact by flaring out so abruptly that it would lose lift and speed several feet above the beach, then drop like a brick.

A large and cumbersome bird can't maneuver as nimbly as its smaller cousins. This makes it even more important for such a bird to follow correct flight procedures. A herring gull at Fisherman's Wharf in San Francisco has to be wary of sudden gusts of wind between the piers and the warehouses, but a few rapid wingbeats and a steeply banked turn will get it out of trouble quickly if something unexpected happens. The brown pelicans that also live here must pay much more attention to the wind. A pelican is a large bird, with a weight of 3 kilograms (7 pounds), a wingspan of 2.20 meters (7 feet), and a wing area of approximately 0.5 square meter (5 square feet). Its wing loading is around 60 newtons per square meter, and its cruising speed is about 12 meters per second (27 miles per hour). Because its wings are enormous, a pelican flies no faster than a herring gull or a homing pigeon, though its weight is comparable to that of the common loon, which must cruise at more than 20 meters per second (45 miles per hour) in order to stay in the air.

While having lunch at Fisherman's Wharf many years ago, I watched a mature pelican that wanted to join a small group of other pelicans waiting for the return of the fishing boats that had gone out to sea that morning. The bird had to make its final approach less than a foot above the mastheads of a dozen fishing boats moored alongside a large warehouse. That necessitated a steep descent in the last 100 feet of the flight. Diving down was out of the question, because that would have made the pelican gain speed just when it needed to lose both speed and altitude. On the first try, everything went wrong. Thirty feet before touchdown the pelican was suddenly blown off course by a vicious crosswind gust— the same kind of mishap that pilots of small planes worry about. The pelican had to shift gears instantly, summoning all the emergency power of its massive wings to perform a steeply banked climbing turn. It avoided a collision with the landing pier by only inches, shifted back to maximum continuous power, made a procedure turn to the right, and flew 300 feet on the downwind leg of the approach pattern in preparation for the next attempt. Then it made a 180° procedure turn to the right to begin final approach, and skimmed the mastheads of the fishing boats in its steep descent toward the pier. The pelican flew as slow and low as it dared. Fortunately there were no further gusts. The bird extended its feet, reduced its speed to the bare minimum, and flared out. It made a beautiful landing, giving no hint of having required all the skill a pelican can muster. It didn't even have to move its feet; it just turned around and joined its friends.

As a young engineer, I once was involved in a "near miss" at Rotterdam Airport. Having recently obtained my private pilot's license, I was practicing takeoffs and landings in a single-engine Saab Safir with the reporting marks PH-UEG ("Echo Golf" in radio communications). In the jargon of pilots this kind of practice is called "touch and go." I was flying in the traffic pattern, on course and on speed, half a mile downwind of the runway threshold, with everything under control. Suddenly, in my earphones, I heard a sharp command: "Echo Golf, turn left *now*." The "now" meant

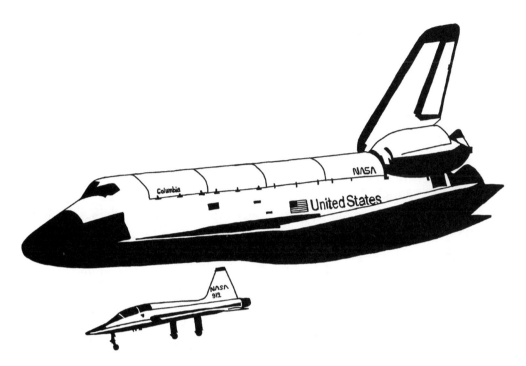

Space Shuttle ($W = 1 \times 10^6$ N, $S = 250$ m^2, $b = 24$ m) and Northrop T-38 ($W = 1.15 \times 10^5$ N, $S = 17.3$ m^2, $b = 8.13$ m).

that the order was to be executed immediately. As I rolled the Safir hard left to begin the required turn, I responded "Turning," confirming that I was following the order without delay. I hadn't noticed it yet, but the traffic controller knew that 15 miles away an airliner had just started its approach. There were still a few minutes to spare, and the man in the control tower evidently thought that there was sufficient room to slip the student pilot on touch-and-go exercises in front of the distant airliner. Even before I arrived in front of the runway, the controller cleared my way: "Echo Golf, cleared to land runway two four." Busy with landing gear, wing flaps, airspeed, carburetor heating, and lots of other details, I continued on "short final," pulling back on the steering column in order to lose speed, extending the wing flaps fully, and closing the throttle all the way. When I was a few hundred feet from the threshold of the runway, with only seconds to go before touchdown, another airliner suddenly started to taxi onto the runway in preparation for takeoff. The pilot had neither seen nor heard me, and he hadn't waited for clearance from traffic control. (One must always request and obtain explicit permission before entering an active runway.) Realizing the danger, I rammed the throttle lever full forward. By doing that I could have killed myself. Airplane piston engines tend to starve from lack of fuel when the throttle is suddenly opened; their carburetors don't have acceleration pumps. The engine hiccupped, and the propeller seemed to stop, but then, thank God, the engine picked up and started to roar. At full power, I just managed to avoid the tail of the airplane that had committed the traffic violation. White around the gills, I flew the traffic circuit once more. Although my touchdown wasn't particularly smooth, I was happy to be on the ground again. Still trembling, I rode back home on my ancient moped. All was well. Afterward I heard my instructor's voice, over and over again: "With the throttle you should be just as considerate as with your girlfriend. Never treat the throttle roughly." Procedures have to be practiced until one can carry them out when rattled.

Brown pelican (*Pelecanus occidentalis*): $W = 30$ N, $S = 0.5$ m^2, $b = 2.20$ m.

I took my ice skates along when I moved to the United States in 1965. That winter, after a couple of nights of hard freeze, I drove up the hills to the small lake behind Whipple Dam in Centre County, Pennsylvania. People were whispering to one another as I tied my skates to my shoes. My antique contraptions, with their leather straps, must have seemed like poor substitutes for decent skates. But the whispers faded when people saw that even the strongest teenager on hockey skates could not keep pace with me.

The transmission of your car and that of your bicycle consist of gears that keep the engine's revolutions and your pedaling rate within limits at high speeds. For the engine this is primarily a matter of fuel economy, but for your legs it is mainly a question of endurance. Muscles at work convert glucose into lactic acid; if the acid cannot be eliminated quickly enough, muscle power drops precipitously. Humans' leg muscles are powerful (a typical athlete's one-hour maximum is about 200 watts), but only if the frequency of the motion remains within limits. A bicycle has a set of gears to make this possible, but on speed skates such complexity is not necessary.

The art of skating produces a fully automatic transmission at absolutely no cost. As you push off, the track of your skate describes a small angle in relation to your direction of travel (figure 11). That angle can be changed. As you increase your forward velocity V, the tracking angle i becomes smaller. You do that automatically. Your legs want to keep the lateral speed w at an acceptable level. This keeps the frequency of your leg movements within limits, thus preventing your muscles from becoming saturated with lactic acid.

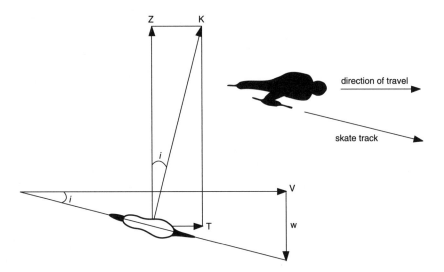

Figure 11 Force and speed for the right-side skating stroke, seen from above. The force K between the ice and the skate is perpendicular to the skate track, because the friction of the skate on the ice is extremely small. The motion of the leg generates not only a large sideways force (Z) but also a forward force (the thrust, T). Because the force triangle has the same proportions as the speed triangle, the ratio T/Z is equal to the ratio w/V between the sideways motion of the skate and the forward motion, V, of the skater.

The geometry of figure 11 has consequences for the triangle of forces, for the relation between the forward speed V and the lateral speed w, and for your energy budget. Since ice skates have hardly any friction in the direction of their motion, the force K that you exert with your leg is perpendicular to the skate track. Because the skate track is at an angle i to the direction of travel, the force K has both a lateral component Z and a forward component T (thrust, as in chapter 2). This is as it should be: without thrust you cannot overcome the aerodynamic resistance.

The force triangle KZT leads to a large lateral force Z and a small thrust T. Because K is perpendicular to the skate track, the angle I between Z and K is identical to the angle i between the direction of travel and the skate track. Hence (this seems trivial, but it is crucial), the proportions within the force triangle are identical to those within the speed triangle. In particular, the ratio between the small thrust T and the large lateral force Z equals the ratio between the low lateral velocity w of the leg and the high forward speed V of the skater:

$$T/Z = w/V. \tag{12}$$

If we write this formula somewhat differently, its elegance is even more striking. Multiplying both sides by V and by Z in order to clear the denominators, we get

$$TV = Zw. \tag{13}$$

The left-hand side of this relation is clearly the power P that is needed for propulsion, just as it is for birds, cars, and airplanes (chapter 2). Power equals force times speed; forward force T times forward speed V equals propulsive power P. But the right-hand side represents some kind of power, too. The lateral force Z exerted by your legs times the lateral speed of the skate strokes equals the power supplied by your legs. Thus, equation 13 states that the work performed by your muscles during the lateral move-ment of your legs is converted *without any loss* into the power needed for propulsion, $P = TV$. This is the basis for your "free gearbox": 100 percent of the work done by the large force Z at the small speed w is converted into the work done by the small force T at the high forward speed V. To increase V, you simply push off harder to increase Z without having to increase the frequency of your leg strokes.

The Art of Flapping

A bird in flapping flight faces essentially the same problem as a skater: because flapping its wings too rapidly as its speed increases will cause a buildup of lactic acid in its muscles, the bird must find a way of flapping that limits the frequency of its wing strokes.

The image of flapping that many people have in their heads is one of ducks and geese rowing through the air with their wings.

Close observation, however, reveals a very different sort of motion. On the downstroke a bird's wings move slightly forward. Rowing with backward strokes would not work. Just imagine moving at 50 miles per hour and having to push your wings back at an even higher rate in order to propel yourself through the air. You would need to maintain a ridiculously high wing-beat rate. You would also spend much more energy than is necessary, because you would generate a lot of unwanted turbulence in the air behind you. Propulsion by paddling is terribly ineffective. If you don't believe me, spend an afternoon on a lake or a canal, alternating between sculling and rowing. That experience, when I was 14 years old, made me an instant convert. Ship designers were converted much earlier: paddle-wheel steamers became obsolete more than a century ago.

The flapping of a bird's wings is like a skater's or a sculler's strokes. The only difference is that the plane of action is rotated 90° (figure 12). The downstroke of the wing must generate both lift and thrust. Because the aerodynamic drag on the wing itself is relatively small, the aerodynamic force K on the wing is almost perpendicular to its direction of motion. As the wing moves down, the force K has not only a vertical component (L), which supplies the lift needed to keep the bird aloft, but also a forward component (T), which provides the required thrust.

The intentional similarity between figures 11 and 12 probably did not escape your attention. The proportions within the force triangle KTL are identical to those within the speed triangle, of which V is the horizontal component and w the vertical component. Therefore, the ratio T/L between thrust and lift equals the ratio w/V between the downward speed of the wing and the forward speed of the bird:

$$T/L = w/V. \tag{14}$$

If we treat this in the same way as equation 12, to ensure that all quantities wind up in the numerator, we obtain

$$TV = Lw. \tag{15}$$

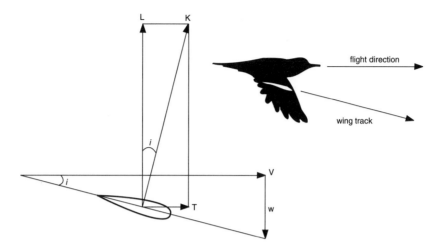

Figure 12 Force and speed for the downstroke of a wing, seen from the side. The aerodynamic force, K, is practically perpendicular to the track of the wing, because the air drag on the wing is quite small. The downstroke generates not only lift (L) but also thrust (T). Because the force and speed triangles have the same proportions, the ratio T/L is equal to the ration w/V between the downward speed of the wing and the forward speed of the bird.

The significance of this is exactly the same as that of equation 13. The product of T and V is the power P required for propulsion. That power is supplied, virtually without loss, by the large force L operating at the small downward speed w. Lift is necessary to overcome gravity and keep the bird in the air, but as the wing moves down the lift force also generates power. This power, the product of L and w, is transmitted in its entirety to the propulsive effort, the product of T and V. Power, which equals force times speed, can apparently be converted at will from large force times small speed into small force times large speed. As long as a bird maintains a small angle between the wing stroke and the direction of flight, it minimizes the loss of energy to the air. That way, it also keeps its wing-beat frequency down.

Are there any data to support this line of thought? Yes, there are. It is useful to introduce a new non-dimensional parameter first. Since wings produce both lift and thrust in the downstroke, it is useful to compare them with propellers. Those are characterized by their "advance ratio," which is the ratio between the airspeed V and the tip speed of the propeller blades. The tip speed is equal to πfd, where d is the propeller diameter and f its turning rate. The advance ratio of a propeller then is $V/\pi fd$. It is made large for high-speed flight, because that minimizes energy losses to the air. But the advance ratio of modern wind turbines, whose task it is to extract as much energy as can be extracted from the air, is made as low as 1:5. Their blades slice through the air like a pineapple cutter.

Flapping wings are like high-speed propellers, and are likely to operate at a high advance ratio. What is the ratio between the flight speed V and the tip speed of its flapping wings? Since there are very few data on tip speeds, we will have to make do with the estimate that the wing-beat amplitude is roughly one-half of the wing-span. The tip speed then becomes equal to the product of the wingspan b and the wing-beat frequency f, and the advance ratio can reliably be estimated as V/fb. Several investigators have measured this non-dimensional parameter and have found values between 2 and 4, with $V/fb = 3$ as the median. This means that the vertical speed of the wingtips is roughly one-third the forward speed of the bird. This number is confirmed by direct tip-speed data on swallows from the wind tunnel at Lund, Sweden. In fact, swallows apparently speed up their downstroke when flying fast, to make sure the tip speed does not get too small. High-speed flight causes more drag, which cannot be overcome with a shallow wing-beat. The barnacle geese featured in chapter 2 are another case in point: with $V = 19$ meters per second, $b = 1.35$ meter, and $f = 5$ beats per second, the advance ratio of their wings, V/fb, is 2.8. When I first drew figure 12, I knew of no data in support of its geometry, but I happened to be right on the nose.

Most birds have little latitude in their wing-beat frequency. It is constrained by muscle physiology. But birds can, and do, vary their wing-beat amplitude. Jackdaws casually switch between gliding and shallow flapping as they scan for food. They are flying slowly, about 8 meters per second, and their drag is low, too. So they need only a little thrust from their wingtips. One-half their regular flight speed, and one-half their regular wing-beat amplitude: the geometry of figure 12 does not change much. But they do need to obtain some lift from their upstrokes, which reduces thrust. Apparently, they don't mind. Swifts don't either: in the Lund wind tunnel their upstrokes cut the effective thrust in half. But if a bird has to accelerate rapidly, as a homing pigeon does immediately after its vertical takeoff or as a pheasant does when trying to escape a fox, all subtleties are ignored. Making the wing-beat amplitude so large that the tips are touching each other at both extremes of each stroke, birds then replace the geometry of figure 12 temporarily by one in which the advance ratio is much smaller. If we take $V/fb = 1$ as a typical geometry during rapid acceleration, then the angle i in figure 12 becomes $45°$, which implies that the air flow over the wingtips is $\sqrt{2}$ faster than the flight speed. The aerodynamic forces, which increase as the square of the speed, thus become twice as large. This does help to generate the sudden burst of thrust required. But quick acceleration comes at a price: the wings now work as fans, transmitting a lot of energy to the jet of air they leave behind.

I believe that birds with advanced aerodynamics can make the advance ratio of their wingbeats larger than average, but I know of only a few data that support this claim. According to Colin Pennycuick, a famous specialist on bird flight, herons fly with $V/fb = 2$, which seems to fit their flight style: these are slow-flying birds with broad wings and a none-too-tight coat of feathers. Kittiwakes (*Larus tridactyla*) and common gulls (*Larus canus*), however, with their slender wings and smooth surfaces, reach $V/fb = 4$. They can do this because their aerodynamic drag is small, requiring only a little thrust for a given weight. But starlings and pigeons at top speed also achieve $V/fb = 4$, although their aerodynamic performance is not nearly as good as that of gulls.

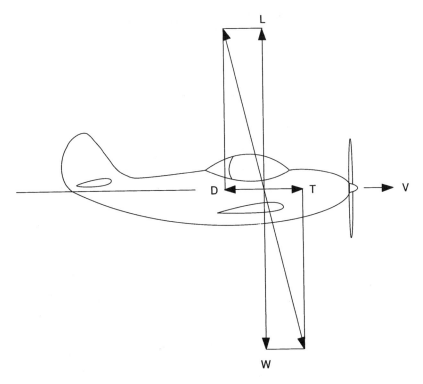

Figure 13   The force balance in horizontal flight. The weight *W* is balanced by the aerodynamic lift *L*, and the aerodynamic drag *D* is overcome by the thrust *T*. The airspeed is *V*, and the product of *T* and *V* is the work that must be performed per second (that is, the power required) to maintain the force balance. Most of the time, *D* is much smaller than *L*. The small *D/L*, the better.

Horizontal Flight

The analogy between skating and flapping flight is useful for understanding how birds and airplanes glide. There too the proportions within the force triangle are equal to those within the speed triangle, and there too both triangles are slender. But first we must recall the frame of reference developed in chapter 2. The quickest way to do that is to look at the force balance in horizontal powered flight (figure 13). The weight *W* is kept in balance by the lift *L*, and the aerodynamic drag *D* is overcome by the thrust *T*. In horizontal flight at constant speed, therefore, $L = W$ and $T = D$.

Power equals force times speed; thus, the relation between thrust T and power P needed to maintain horizontal flight is given by $P = TV$, just as in skating or flapping. We can better judge the energy requirements of anything that flies by computing the specific energy consumption $E = P/WV$, which was introduced in chapter 2. The force balance in figure 13 allows us to transform E in various ways. With the aid of $P = TV$ we obtain

$$E = P/WV = T/W = D/L. \tag{16}$$

The ratio between drag and lift determines the specific energy requirements. Tucker's parakeet achieved a minimum of 0.22 at a speed of 11 meters per second (25 miles per hour). For parawings and most small birds, the value of D/L is comparable. Seagulls ($D/L = 0.07$), jetliners ($D/L = 0.06$), and albatrosses ($D/L = 0.05$) are much more economical. Extremely low values of D/L are obtained by state-of-the-art soaring planes, which easily attain $D/L = 0.025$ and in some cases even $D/L < 0.02$. (All these values are minima; at speeds above or below the optimum value, the specific energy consumption is higher.)

If you want to save energy in flight, you have to minimize D/L. In flight performance the ratio between drag and lift is a measure of aerodynamic quality. It is inconvenient, however, that this number decreases as the aerodynamic quality is improved. It would be better if we could arrange matters so that the quality number increases as the energy needs decrease. And that is quite easy: just turn D/L upside down. The quantity L/D, to which French aeronautical engineers have given the beautiful and appropriate name *finesse*, has the very properties we desire:

$$L/D = 1/E = F. \tag{17}$$

It is a pity that aeronautical engineers in most other countries have given F such unimaginative names. Dutch and German engineers use the equivalent of "glide number" (correct but dull). English-speaking engineers use "glide ratio" (equally correct and equally dull).

The best finesse a grass parakeet achieves is $1/0.22 = 4.5$. The finesse of a wandering albatross is about 25, and that of a Boeing 747 at cruising speed is about 16. The 747 can do somewhat better ($F = 18$) when it flies slower, but its engines are less efficient at lower speeds. Advanced sailplanes achieve $F = 40$ with ease; some reach $F = 60$. The finesse of a bird or an airplane can be enhanced by slender wings and a smooth, streamlined body.

The Subtleties of Gliding

In the absence of thrust, an airplane cannot maintain the balance of forces required for horizontal flight; it will inevitably lose altitude. When an airplane descends, a new balance is obtained, the component of the weight W directed along the glide path becoming equal to the drag D (figure 14). As with a bicyclist freewheeling downhill, gravity takes over. The lift L and the drag D constitute the aerodynamic force K that balances the weight W. And, just as in skating and wing flapping, the force triangle is slender; in most cases D is much smaller than L.

The force and speed triangles in figure 14 have the same proportions. Hence, the ratio between the rate of descent w and the airspeed V equals the ratio between the drag D and the weight W:

$$w/V = D/W. \tag{18}$$

As with skate strokes and wingbeats, this can be put in a form that clarifies the energy budget of gliding. Multiplying equation 18 both by W and by V, we obtain

$$Ww = DV. \tag{19}$$

The power $P = DV$ required to overcome the aerodynamic drag is apparently supplied by the large force W acting at the small downward speed w. Power equals force times speed—in this case, the force of gravity times the rate of descent.

The proportionality between the force and speed triangles also provides useful information on the distance that can be covered in

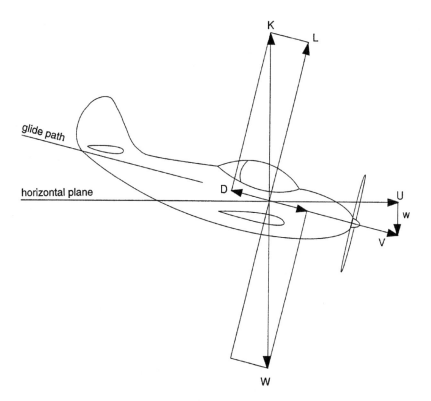

Figure 14 The force balance in gliding flight. With the throttle closed, the thrust equals zero. The drag D now must be overcome by the component of the weight W that is directed along the glide path. Again the force triangle has the same proportions as the speed triangle; hence $D/W = w/V$, where w is the rate of descent and V is the airspeed, now along the glide path. The work performed per second by the weight W is equal to wW, but also to DV.

a glide. If we call the horizontal component of the airspeed U, the ratio between it and the rate of descent w must be equal to the ratio between the lift L and the drag D. But the ratio L/D is the finesse, F. What we discover is that the finesse (and only the finesse; no other quantity is involved) determines how many meters a gliding bird or plane can travel for each meter of descent:

$$F = L/D = U/w. \tag{20}$$

Black-browed albatross (*Diomeda melanophris*): $W = 38$ N, $S = 0.32$ m^2, $b = 2.20$ m.

It is for this reason that aeronautical engineers speak of glide ratio when they talk about the finesse of an airplane. A Boeing 777 has a glide ratio of almost 20. Should both its engines fail at an altitude of 10 kilometers (33,000 feet), the plane can remain airborne for another 200 kilometers (125 miles). An airliner flying to Amsterdam from the United States starts its descent over the British Isles, well before crossing the North Sea. For the same reason, almost half of the flying time on short hops (such as from Chicago to Detroit) consists of descending flight. A jet can easily reach an alternate airport in case of engine failure over Europe or the continental United States; at almost any point there are several airports within 100 miles. But a failure of *both* engines just after takeoff requires quick decisions. The pilot who safely landed his crippled airliner on the Hudson River on January 15, 2009, acted with superb professional judgment. Deservedly, he became the hero of the "miracle on the Hudson."

A glider with $F = 40$ can travel 40 feet per foot of altitude lost. At a distance of 4,000 feet from its landing spot, the glider needs to be only 100 feet above the ground. Since that is barely above the treetops, this cannot be regarded as a safe approach procedure. A pilot must be able to see where he or she is going, and the view should not be obstructed by trees or buildings. For this reason, all sailplanes are fitted with air brakes (also known as spoilers). With spoilers extended, a sailplane can descend steeply. During the final approach, just before touchdown, $F = 10$ is more than adequate. (The spoilers on some automobiles are not meant to increase drag but to spoil unwanted aerodynamic lift and keep the wheels in contact with the road at high speeds. But those spoilers are not effective at legal highway speeds; the aerodynamic forces are simply not large enough.)

In a glide, the power $P = DV$ needed to overcome the drag D is supplied by gravity: $P = Ww$ (equation 19). But this means that the rate of descent w can be used as a measure for the engine or muscle power that a bird or a plane must have available to stay aloft:

$$w = P/W. \tag{21}$$

The rate of descent in gliding is equal to the ratio between the power P needed to maintain horizontal flight and the weight W (the "power loading," P/W). The significance of this can be grasped easily by recalling some data from chapter 2. On long flights the pectoral muscles of a bird supply about 100 watts per kilogram of muscle mass. But the flight muscles account for about 20 percent of a bird's mass. The flight muscles therefore supply about 20 watts per kilogram of body mass, which amounts to 2 watts per newton of overall weight. As table 2 shows, watts per newton are meters per second. Equation 21 states that birds with rates of descent greater than 2 meters per second (400 feet per minute) do not possess muscles strong enough to keep them aloft for any length of time.

The Great Gliding Diagram

According to equation 21, we must look not only at the finesse but also at the rate of descent when we want to judge the performance of birds and airplanes. We can do so in an orderly way by plotting rate of descent against airspeed (figure 15). In this figure—again a proportional diagram, so that the relative proportions remain the

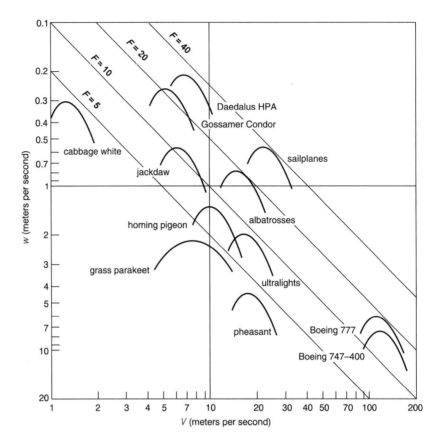

Figure 15 The Great Gliding Diagram. Airspeed is plotted on the horizontal axis. Rate of descent is plotted along the vertical axis, downward. The diagonals running from top left to bottom right are lines of constant finesse. The practical soaring limit, 1 meter per second, is indicated by the horizontal line.

same all over—the performance characteristics of insects, birds, and planes can be compared in a straightforward way.

There is a great deal to be learned from figure 15. First let's look at gliding birds. A pheasant descends at more than 4 meters per second—twice the maximum we just calculated. A pheasant cannot fly for more than a minute or so; a quick dash to escape a fox or a hunting dog is all it can manage. A parakeet, with a rate of descent of 2 meters per second, is capable of continuous flight, but without any power to spare. A larger bird with the same poor value of F (a pheasant, for example) would not have lasted long in Vance Tucker's wind tunnel. But a swift (*Apus apus*) has plenty power to spare for aerobatics. I didn't include a curve for swifts in figure 15, because it overlaps the curve for jackdaws, about which much more will be said a bit later. The continuous power rating of a swift is 2 watts per newton, but it needs only 0.6 watt per newton to maintain horizontal flight. Hence, it has 1.4 watt per newton to spare. This means that it can climb 1.4 meter per second (roughly 300 feet per minute) without much effort. And if it really wants to exert itself in a brief climb, its muscles may be able to produce 4 times the normal rate, or 8 watts per newton. With only 0.6 watt per newton needed to maintain altitude, the remaining power can be used for ascending 1,500 feet per minute. That is a faster rate of climb than that of a Beech Bonanza. The top speed of swifts is equally impressive. In a dash, with 4 watts per newton and a finesse of only 6, swifts have been observed to achieve 20 meters per second (45 miles per hour). The impression of speed is heightened because the human eye judges the speed of a flying object by the object's size. At its maximum cruising speed, a swift travels 40 wingspans per second. A cruising 747 manages four wingspans per second, a 747 on final approach less than two.

Homing pigeons (*Columba livia*) occupy center stage in figure 15. They have been used in many wind-tunnel studies over the years, but not all the data obtained that way are mutually consistent. For one, gliding speeds obtained in wind tunnels (around 10 meters per second) are much lower than the speeds reported by owners of racing pigeons. Pigeon fanciers (as they are called) quote numbers up to 20 meters per second (45 miles per hour). That number agrees with radar data, but not with the gliding performance given in figure 15. If we extrapolate that curve to 20 meters per second, we get a power need of about 8 watts per newton. Can a racing pigeon achieve that? Its chest muscles produce 2 watts per newton of body weight, not nearly enough to support the observed racing speed (a horizontal line at 2 meters per second of descent

Whooping crane (*Grus americana*): $W = 68$ N, $S = 0.6$ m^2, $b = 2.2$ m.

crosses the pigeon curve at a flight speed of 14 meters per second). This discrepancy can be overcome by realizing that flapping wings are two-stroke engines, in which the downstroke has to produce an amount of lift that is equal to twice the weight. The consequence of this train of thought is that the curve in figure 15 has to be shifted 40 percent to the right, down along the diagonals, before it can be used for flapping flight. When you do that and repeat the calculation, you will find that the problem resolves itself. (I will not give the numbers for pigeons, but within a few pages I will do these sums for jackdaws, the birds that in 2002 overturned my previous thinking.)

In general, larger birds must fly faster. With the same aerodynamic quality, which is to say at the same value of finesse, birds therefore slide along the diagonals toward the lower right corner in figure 15 as they become heavier. Hence, they require proportionally more power to remain airborne. Inevitably there comes a moment when the flight muscles are no longer strong enough. If we set this limit, as we did above, at 2 watts per newton, corresponding to a rate of descent of 2 meters per second, and if we take $F = 12$ as the greatest attainable practical value for the glide ratio of large birds, then the maximum achievable cruising speed becomes $2 \times 12 = 24$ meters per second (54 miles per hour). Thanks to their extremely slender wings, albatrosses achieve $F = 20$ or more (for the largest species, the wandering albatross *Diomedea exulans*, I estimate $F = 25$), but continuous flapping flight is beyond their capabilities. A cruising speed of 24 meters per second demands a wing loading of 220 newtons per square meter. For a bird of average proportions (i.e., one that stays in the vicinity of the main diagonal in figure 2), the corresponding weight is about 100 newtons (10 kilograms, 22 pounds). Birds that wish to maintain flapping flight for extended periods should not exceed this limit. Those that do need oversize wings to reduce their power requirements. But oversize wings are ill suited for continuous flapping. It is no coincidence that most of the very large birds specialize in soaring.

Piston-engine airplanes have a similar upper limit. An aviation gasoline engine weighs roughly a kilogram per kilowatt of takeoff power, and about half of that power is available in cruising flight. The specific cruising power, therefore, is 500 watts per kilogram, or 50 watts per newton. A practical limit for the weight of the engine(s) is 20 percent of the plane's total weight. This gives $P/W = 10$ watts per newton of total weight. The corresponding rate of descent (equation 21) is 10 meters per second. Again we choose $F = 12$ as a representative value of the finesse, which makes the cruising speed 120 meters per second (270 miles per hour). According to figure 2, the wing loading should then be about 5,000 newtons per square meter. An average airplane with that wing loading weighs about a million newtons, or 100 tons. As Howard Hughes discovered with his Spruce Goose, a piston-engine airplane bigger than that (for example, one the size of a 747) is dangerously underpowered. Aircraft manufacturers switched to jet engines as soon as they could because jet engines have much better power-to-weight ratios than piston engines.

If a bird or a plane wants to stay aloft effortlessly for extended periods, its rate of descent must be less than the strength of the updrafts in the air along mountain ridges or in convective thermals. A practical limit is 1 meter per second. (See figure 15.) Parakeets, chickens, partridges, and pheasants do not need to bother about learning to soar; they descend far too quickly. Modern gliders achieve a rate of descent as small as 0.6 meter per second (120 feet per minute). They can do this because they have extremely slender wings and carefully maintained, smoothly polished skins. Albatrosses are similarly specialized: their minimum rate of descent is less than 0.8 meter per second. Great soaring birds of prey, such as the golden eagle and the California condor, have solved the problem in a different way: they have developed relatively large and broad wings, which give them low cruising speeds. Their rates of descent are within the soaring limit of 1 meter per second, though they do not come anywhere near the albatross in finesse.

Butterflies can soar without worrying much about aerodynamic quality ($F = 4$). Their wing loadings are so low that they can descend at a rate of 0.3 meter per second. But the real mavericks in figure 15 are human-powered airplanes, represented in figure 15 by the first of the species, the *Gossamer Condor* (1975), and the last, the *Daedalus* (1988). With the pilot on board, human-powered aircraft weigh about 100 kilograms, yet their rate of descent is less than that of a 0.15-gram cabbage white.

Trailing Vortices and Induced Drag

It seems quite strange that flying should be uneconomical at low speeds. In all other forms of locomotion (including swimming, bicycling, and driving automobiles) the drag increases as the square of the speed, so twice as fast means 4 times as much drag. But everything that flies can choose an optimum speed, and that optimum need not be unfavorable if it happens to occur at high speeds. Though a Boeing 747 flies much faster than a swift, it has

Painted Lady (*Vanessa cardui*).

about the same finesse. In other words, as a fraction of its weight the drag of a 747 is no larger than that of a swift—in fact, it is a bit lower: a 747-400 has $D/L = \frac{1}{17}$, while a gliding swift has $D/L = \frac{1}{13}$. Evidently there must be a drag component that decreases as the speed goes up. It is helpful here to take another look at figure 7, where the drag of Tucker's flapping parakeet is plotted against speed. It is evident that the drag of a bird comes down in a hurry when the speed goes up. It does not begin to increase until much higher speeds are reached. This is no small matter. I muddled through it in the first edition of this book because I did not dare to cope with the intellectual hurdles involved. (To handle this subject elegantly, one must use differential analysis of continuous vector fields.) I was not pleased with myself, and neither were colleagues and reviewers. This time I will do better.

The best way to explain the strange behavior of drag as speed increases is to employ what is called the "vortex theory of lift and drag." I will do so at a leisurely pace, because it is not easy going, though it is immensely rewarding. If you're not eager to follow me in this, you may skip forward to equation 28.

In freshman physics or electrical engineering courses, many of us have learned about the electromotive force, the force that makes electric motors spin. When an electrical current flows through a

A Pilatus Turbo Porter, the perfect airplane for rough jobs and short airstrips. It has fixed landing gear and a powerful turboprop engine: $W = 27.7$ kN, $S = 30$ m^2, $b = 16$ m, $P = 410$ kW (550 hp).

copper wire that is exposed to a magnetic field, a force is exerted on the wire that is at right angles both to the wire and to the magnetic field. The strength $\boldsymbol{F}$ of the electromotive force is proportional both to the strength of the magnetic field B and the strength of the electric current J. In shorthand: $\boldsymbol{F} = \boldsymbol{J} \times \boldsymbol{B}$. Electrical engineers prefer to understand this force as a consequence of the way the magnetic field is distorted by the "induced magnetic field" that is associated with the electric current through the copper. The induced magnetic field wraps itself around the wire in closed loops. Adding these to the imposed magnetic field, you will find that the distorted field has a pronounced convex curvature on one side of the wire.

Ludwig Prandtl, the German grandfather of aerodynamics, was thoroughly familiar with these concepts. About 100 years ago, he realized that the flow field around a wing can be interpreted in a similar way. A wing deforms the flow field around it and, in doing so, creates a force that is perpendicular both to the air flow and to the "corkscrew" motion of the induced flow that it creates by moving through the air. That force, the lift L, has a strength that is proportional both to the airspeed and to the amount of flow distortion generated by the wing. The only difference between this case and that of the electromotive force is that a current-carrying wire is pushed away from the convex side of the field distortion, whereas a wing is pulled toward the convex side.

The aerodynamic counterpart of $\boldsymbol{F} = \boldsymbol{J} \times B$ now becomes

$$L/b = \rho V \times \Gamma. \tag{22}$$

Here L is the lift and Γ is the "circulation" around the wing. The circulation is expressed in meters of circumference times meters per second (that is, meters squared per second). The circulation is the product of the length of a contour around the wing and the tangential speeds prevailing there. I use the lift per unit span, L/b, in equation 22, because the electromotive force is also expressed per unit length. The air density ρ is needed to make sure that the force is larger in a denser medium and to give equation 22 consistent units.

Air circulating around a wing—does that agree with the discussion of lift in chapter 1? Let us first check whether equation 22 is consistent with the idea that a wing gives the airflow around it a downward impulse. At the trailing edge, we found that the downward component of the airspeed is proportional to αV, where α is the angle of attack. Imagine that this is made possible by an imaginary vortex surrounding the cross-section of the wing. The strength of this "bound vortex," which we now dare to equate to Γ, must be proportional to αV times the wing chord c, or else the required downward component of the airspeed is not obtained. With Γ now proportional to αVc, equation 22 turns into

$$L/b = n\rho V \times \alpha Vc. \tag{23}$$

Here n is a non-dimensional coefficient that is irrelevant to this train of thought. If we multiply both sides by the wingspan b, and replace L by the weight W, this converts to

$$W = n\alpha\rho V^2 bc = n\alpha\rho V^2 S. \tag{24}$$

The contrails of a Boeing 747 flying overhead wrap themselves around the trailing vortices. The contrails from the inboard engines temporarily fan out farther than those of the outboard engines.

Obviously we are on the right track, because equation 24 agrees with equation 1 in chapter 1.

Prandtl knew that electric motors have "winding losses" or "end losses" caused by the finite length of the wires in the magnetic field. He also knew how to calculate these losses: with the formalism known as the Biot-Savart equations. How should he proceed? He realized that the circulation around a wing cannot stop at the wing's tips. An electrical current, the analogue of the circulation, has the same property: it cannot stop cold, but has to continue in the parts of the winding that are outside the magnetic field. So Prandtl concluded that the circulation around the wing must come off the wingtips as a pair of "tip vortices." The circulation has no choice but to continue without any loss of strength as two long rotating braids in the air behind the wing. These two trailing vortices mark the path that a bird or a plane has carved in the air. If you don't believe me, just go outside on a clear day and watch the contrails of a high-flying jetliner. Since it is very cold at 30,000 feet, the moist exhaust gases from the engines soon condense and freeze into tiny crystals. The exhaust plumes made visible this way are subsequently sucked up by the tip vortices. Figure 16 and other illustrations on these pages give an impression of this beautiful phenomenon.

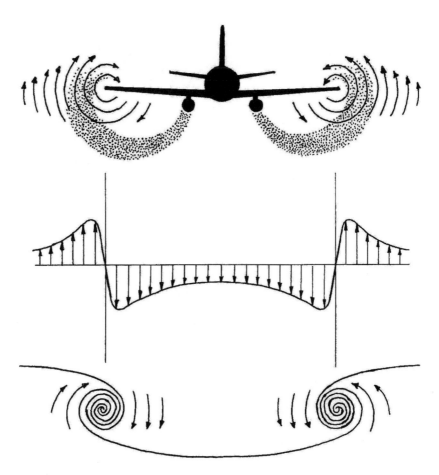

Figure 16 Top: A jetliner seen from behind, with tip vortices and the formation of exhaust contrails shown. Note that the distance between the contrails far behind the plane is
much larger than the distance between the engines. Center: A representation of the
vertical velocity field behind the wing, including the upwash beyond the tips, which
is exploited by wingmen and geese flying in formation. Bottom: A sketch of the
channel in the air caused by the airplane.

Trailing vortices are dangerous! Take the case of a Boeing 747-400 on final approach. If we substitute W for L in equation 22, it converts to

$$\Gamma = W/(\rho Vb). \tag{25}$$

What are the relevant numbers? At the end of a long trip, the weight of a 747 is down to 300 tons, or 3 million newtons. The approach speed is about 70 meters per second, the air density 1.24 kilogram per cubic meter, and the wingspan 65 meters. With these numbers, Γ computes as about 530 meters squared per second. Thirty meters away from the core of one of these whirlwinds, the circumference is about 200 meters long. Circulation equals circumference times tangential speed, so the latter is more than 2.5 meters per second (almost 9 feet per second, or 500 feet per minute) 30 meters out. Don't get near if you are piloting a small plane! In fact, the separation rules for air traffic are designed to avoid these risks.

Trailing vortices force each other down. They are about one wingspan apart, and they induce in each other a downdraft equal to $\Gamma/2\pi b$. For our 747 on final approach, this is a bit more than 1 meter per second, more than 200 feet per minute. The downwash on the centerline is 2.6 meters per second, or 500 feet per minute. In aviation parlance, this is called "wake turbulence." Turbulent or not, it is wise to keep a very safe distance away from these trenches, even if the plane in front of you is a lot smaller than a 747.

The message of this story is not all bad and dangerous, though. If you dare to fly in the upwash a little away from the wingtip of the lead pilot up front, you can throttle back a little. This is why geese typically fly in V-shaped formations. The advantage to the flock as a whole is spread over all its members. Every so often, the lead goose veers off to rejoin the end of the pack, where it can relax a little.

The downdraft created by the interaction between the two tip vortices also produces downwash in the flow just in front of the wing. This means that a plane or a bird has to cope with the downwash it is continuously creating. That requires an additional amount of power, just as climbing a hill does.

How much power is needed to climb this hill? In front of the wing, the local downdraft produced by the tip vortices is given by

$$w = \Gamma/b = W/\rho V b^2. \tag{26}$$

Lifting the weight W up this slope requires that work be performed at a rate given by Ww joules per second (Ww watts):

$$Ww = P_i = W^2/\rho V b^2, \tag{27}$$

where P_i is called "induced power". Induced power is the rate at which the trailing vortices are supplied with kinetic energy. This is how the additional effort mentioned a moment ago is spent. To obtain the total cost of flying, the work that has to be done to create trailing vortices must be added to the work that has to be done to overcome friction. It would have been logical to speak of "wingtip losses," but that is not the way it is phrased in aeronautical jargon. Instead, equation 27 is reformulated as if the rate of working against the self-induced downdraft were an actual contribution to the total aerodynamic drag. Power equals drag times speed, so drag equals power divided by speed. The "induced drag" D_i can therefore be written as

$$D_i = P_i/V = W^2/\rho V^2 b^2. \tag{28}$$

This is no minor matter. The square of the airspeed appears in the denominator, not the numerator. When the airspeed increases, the induced drag decreases in a hurry. When an airplane flies twice as fast as its design speed, its induced drag decreases by a factor of 4. The other side of this coin is devastating, however: when the speed is halved, the induced drag becomes 4 times as large. This is why everything that flies is uneconomical at low speeds, and this is why the most economical speeds of birds and airplanes are high relative to those of other ways of locomotion and transportation. Animals that move on land are supported by the ground; animals that fly have to support themselves by speeding—and the faster the better, because sufficient lift to overcome gravity then is obtained without causing large disturbances in the air passing by.

Equation 28 has another surprise in store. It is not only the square of the speed that appears in the denominator; the square of the wingspan b is there too. When you manage to double the wingspan, you are awarded by a fourfold reduction in induced drag. (Birds could also generate the lift required by surfing on their bellies, which is how human skydivers do it. But that generates outrageously high induced drag, giving your flight an abominably low finesse. If you want to fly economically, your wings must be slender.)

The slenderness of wings is the ratio between length and width. But the width of most wings does not remain constant along the span; it usually tapers off toward the tips. For this reason, aeronautical engineers use the ratio b^2/S. This ratio, called the "aspect ratio" in the English technical literature, is assigned the symbol A. Dutch aerodynamicists use the equivalent of 'slenderness' instead, conjuring up images of shapeliness, grace, elegance, and refinement. That is most appropriate, as we shall see in a moment. Using $A = b^2/S$ and dividing equation 28 by W, we obtain

$$D_i/W = W/\rho V^2 SA. \tag{29}$$

The right-hand side of this equation contains the same $\rho V^2 S$ that occurs in equation 24. Using that opportunity, we can write the induced drag as

$$D_i/W = n\alpha/A. \tag{30}$$

We are on the right track here. The induced drag decreases with decreasing angle of attack α and with increasing aspect ratio. That makes sense. The angle of attack decreases when one flies faster, and faster flight means less energy loss to the tip vortices. The advantages of a high-aspect-ratio wing are obvious: if the distance between the wingtips is large, the energy lost in the tip vortices is small.

If we were to go on this way, we might come close to convincing ourselves that wings should always be as slender as they can be. But that isn't true. Aerodynamic friction counterbalances induced

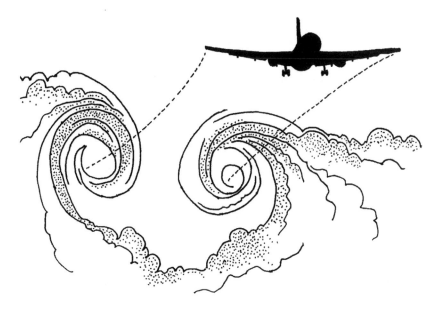

When an airplane has just passed the top of a cumulus cloud, its tip vortices cause quite a stir.

drag. Induced drag decreases as the square of the airspeed, but frictional drag increases at the same rate, as was explained in chapter 2. The total drag is the sum of frictional drag and induced drag; in the simple theory used here, it has a minimum when the frictional drag equals the induced drag. On that basis, a few more calculations, found in the professional literature but not reproduced here, give the following results. At the airspeed that maximizes the finesse L/D, we obtain

$$L/D = \sqrt{(A/f)}, \tag{31}$$

$$\alpha = 10\sqrt{(fA)}. \tag{32}$$

Here f is the friction factor, a non-dimensional representation of the frictional drag. The factor 10 in equation 32 has been chosen so that the angle of attack, α, can be measured in degrees. Keep in mind that when α goes up speed comes down, and vice versa.

A Boeing 777 has a very good finesse: $L/D = 20$. Relative to most airliners, its wings are slender: $A = 8.7$. Putting these numbers into equation 31, we obtain the friction factor $f = 0.022$. Using this in equation 32 and recalling that $A = 8.7$, we obtain $\alpha = 4.5°$. Or take a gliding jackdaw, with $A = 6.2$ and $L/D = 12.5$, as measured in Lund. Its numbers are $f = 0.040$ and $\alpha = 5.0°$. How about a wandering albatross? With $A = 19$ and $L/D = F = 25$, we find $f = 0.030$ and $\alpha = 7.5°$. Open-class sailplanes have $A = 38$ and $F = 60$, so that $f = 0.011$ and $\alpha = 6.3°$.

Equations 31 and 32 make perfectly clear how the case of friction versus aspect ratio should be argued. Equation 31 shows that low friction and high aspect ratio contribute equally to the glide ratio, and equation 32 supports that by saying that we shouldn't fool with the angle of attack if a high glide ratio is our goal. In fact, equation 32 supports the crude but robust choice made in chapter 1: economical cruising flight is achieved when the angle of attack is kept close to 6°.

What goes wrong if you don't follow this wisdom? Suppose you want to give an airplane high-aspect-ratio wings, but are willing to cope with a high friction factor (perhaps because you refuse to invest in retractable landing gear)? Equation 31 shows that this reduces L/D, which was to be expected. But equation 32 shows that α has to increase, perhaps to a point where your airplane is stalling because it has lost too much speed. With more frictional drag, you have no choice but to slow down. The wings of your plane will now keep you in the air only if you pull back on the control stick. And since you have to fly slower, you might just as well choose larger but stubbier wings. In short: if f is larger than usual, A might just a well be smaller than you first had in mind. For house sparrows and hang gliders, smooth streamlining does not have high priority. Their wings must be able to withstand frequent folding and considerable abuse, since accidental collisions with obstacles occur all too often. Conversely, if you want to get the best of both worlds, you have to aim for the smallest feasible friction factor and the largest glide ratio that you can manage. The word 'slenderness' thus acquires a deeper meaning. If you want to achieve a high value for the finesse, you will have to give your plane elegant, slender wings, but you also will have to improve its streamlining: sleek curves, smooth surfaces, absolutely no loose or poorly fitting parts, no dangling legs, everything flush and tight.

It is useful to continue this train of thought with a few more numbers. The major difference between the Boeing 747-400 and its predecessors, the 747-200 and the 747-300, is the increase in wingspan from 60 to 65 meters. Correspondingly, the aspect ratio A increased from 7.0 to 7.8, about 11 percent. Assuming that the friction factor didn't change from one model to the next, this gives a more than 5 percent increase in the glide ratio. Boeing's engineers played it safe: they claimed a 3 percent decrease in fuel consumption. And in designing the 777, they bet on the same horse by increasing the aspect ratio even more ($A = 8.7$), thus adding another 5 percent to the finesse. Boeing discarded the once-fashionable habit of increasing the effective span of a wing by fencing it off with winglets, such as those that grace the tips of the 747-400 and several recent versions of the 737. If you want a higher aspect ratio, you might just as well increase the length of the wing spars. Structural engineers prefer that option.

With these deliberations in mind, it is instructive to study the values of F and A for various birds. A number of examples have been collected in table 5. Many of the numbers for F given in this table are quite different from those in the first edition of this book. For example, I now list crows as having a finesse of 10, not 5. My estimates for the performance of other birds have changed considerably, too. I have assigned $F = 15$ to herring gulls because their wings are so much narrower than those of a jackdaw, which reaches $F = 12$ in the wind tunnel. Recent wind-tunnel measurements of swifts and barn swallows determine the numbers given for them, and the observed migration performance of godwits, knots, and plovers necessitates that their finesse be estimated as $F = 14$. The values given for large soaring birds of prey (raptors, to the experts) are based on their observed soaring performance.

Table 5 Aspect ratio A and finesse F for various birds and airplanes. The values for A have been calculated, those for F have been measured or estimated.

	W (N)	S (m^2)	b (m)	A	F
Magnolia warbler	0.09	0.007	0.20	6	4
House wren	0.11	0.005	0.17	6	4
Barn swallow	0.17	0.012	0.31	8	8
Chimney swift	0.17	0.010	0.32	10	10
Tree swallow	0.20	0.013	0.32	8	8
Orchard oriole	0.23	0.010	0.24	6	4
House sparrow	0.28	0.009	0.23	6	4
Swift	0.36	0.016	0.42	11	13
American robin	0.82	0.024	0.38	6	7
Purple martin	0.43	0.019	0.41	9	9
Starling	0.83	0.024	0.38	6	7
Blue jay	0.89	0.024	0.38	6	7
Common tern	1.2	0.056	0.83	12	14
Red knot	1.3	0.029	0.50	9	14
Merlin	1.4	0.044	0.60	8	9
Hobby	1.7	0.056	0.75	10	10
Kestrel	1.8	0.060	0.74	9	10
Jackdaw	2.4	0.068	0.65	6	12
Montagu's harrier	2.4	0.130	1.10	9	11
Rock dove	2.9	0.075	0.80	8.5	8
Bar-tailed godwit	3.2	0.052	0.73	10	14
Hen harrier (male)	3.3	0.140	1.00	7	10
Cooper's hawk	4.3	0.090	0.71	5.6	9
Royal tern	4.7	0.108	1.15	12	14
Hen harrier (female)	4.7	0.176	1.15	7.5	11
Wood pigeon	4.9	0.082	0.75	7	8
Barn owl	5.0	0.168	1.12	7.5	8
Carrion crow	5.7	0.138	0.91	6	10

Table 5 (continued)

	W (N)	S (m²)	b (m)	A	F
Marsh harrier	6.5	0.204	1.16	7	10
Goshawk (male)	7.0	0.170	0.97	5.5	9
Peregrine falcon	7.9	0.126	1.02	8	10
Red-shouldered hawk	8.0	0.166	1.02	6	11
Common buzzard	8.9	0.269	1.24	6	11
Herring gull	11	0.197	1.34	9	14
Red-tailed hawk	11	0.209	1.22	7	11
Goshawk (female)	12	0.240	1.15	5.5	10
Pheasant	12	0.088	0.72	6	4
Brent goose	13	0.113	1.01	9	12
Osprey (male)	13	0.26	1.45	8	12
Turkey vulture	15	0.44	1.75	7	11
Barnacle goose	17	0.115	1.08	10	12
Osprey (female)	20	0.30	1.60	8.5	13
Black vulture	21	0.33	1.38	6	10
Cormorant	22	0.224	1.40	9	10
Sooty albatross	28	0.34	2.18	14	20
White stork	34	0.50	2.00	8	10
Black-browed albatross	38	0.36	2.16	13	20
Golden eagle	41	0.60	2.03	7	14
Bald eagle	47	0.76	2.24	6.6	15
White-tailed eagle	50	0.72	2.10	6	14
Canada goose	57	0.28	1.70	10	14
Griffon vulture	70	1.00	2.60	7	15
Wandering albatross	85	0.62	3.40	19	25
Mute swan	106	0.65	2.30	8	10
Gossamer Condor	1,000	70	29	12	20
Daedalus	1,000	31	34	37	38
Hang glider	1,000	15	10	7	8

Table 5 (continued)

	W (N)	S (m²)	b (m)	A	F
Parawing	1,000	25	8	3	4
Powered parawing	1,700	35	10	3	4
Ultralight	2,500	15	10	7	8
Standard-class sailplane	3,500	10.5	15	21	40
Icaré solar plane	3,600	25	25	25	35
Open-class sailplane	5,500	16	25	39	60
Boeing 737-900	850,000	125	35	10	15
Boeing 777-300	3,510,000	435	65	10	20
Boeing 747-400	3,950,000	524	65	8.1	16
Airbus A380	5,600,000	845	80	7.5	16

Several smaller birds of prey have flown in wind tunnels; I base my estimate for buzzards ($F = 11$) on those data. Wherever it seemed necessary, I took into account that slender wings lead to better glide ratios. Montagu's harrier (*Circus pygargus*) has very slender wings, but I did not dare to assign it a value of F better than the measured value for a jackdaw, because no other raptor of that size comes even close. I hope a biologist with a wind tunnel will attempt to find out if the number I give is too conservative.

There is a significant difference between standard-class and open-class gliders. For good reasons, the standard class has a prescribed wingspan: 15 meters. If the span were left to the discretion of the designers, everyone capable of building a wing of greater span would be able to achieve a higher finesse (that is, a better glide ratio). A rigid limit on the wingspan amounts to a firm upper limit on A, and thus the standard class needs no handicap rules. Like an albatross, a standard-class glider has an aspect ratio of about 20.

In the open class, designers attempt to achieve extremely high values of A. This is no minor undertaking. Aspect ratios as high as 40 make sense only when the skin of the wings, the tail, and the body is extremely smooth, with every seam or crack securely taped. Besides that, the wing spars must withstand enormous bending forces with little flexing, and this makes a sailplane much heavier. The empty weight of a standard-class sailplane is about 250 kilograms (550 pounds); that of an open-class sailplane is about 450 kilograms. Nevertheless, it is tempting to increase the aspect ratio of the wings.

Buzzards, eagles, vultures, harriers, and various other birds of prey spend their days soaring around in thermals. A relatively low wing loading keeps a bird's airspeed down, enabling it to achieve a low rate of descent (less than a meter per second). A typical aspect ratio for such a bird is 7. Nevertheless, the finesse of these birds is higher than expected: $F = 10$. Some investigators have speculated that fully spread primary feathers (the strong quills on the tip of each wing), with wide gaps between them, produce an effect similar to that of increased wingspan (figure 17). A final verdict has not yet been reached.

The Jackdaw in Sweden: Gliding and Flapping

A young female jackdaw (*Corvus monedula*) demonstrated superior gliding performance in the Lund wind tunnel. This forced me to make radical revisions in my thoughts on flapping and gliding. Two things stood out: its finesse (glide ratio) was much better than I had assumed for many years, and its best gliding speed was much lower than any of the travel speeds I had seen in the research literature.

As far as the gliding is concerned, I could have known better if I had just been more attentive. I once had lunch in a restaurant embedded in the cliffs on the North Cape at Lanzarote, one of the Canary Islands. These cliffs are exposed to the northeasterly trade winds, which make them an ideal location for soaring. Brilliant,

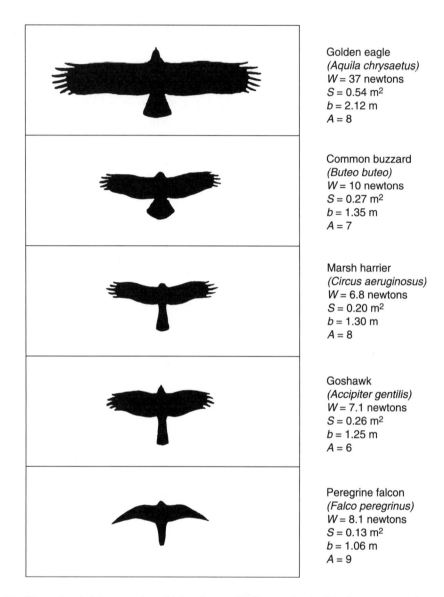

Figure 17 Dimensional data on various birds of prey. All the species in this figure except the peregrine falcon (a high-speed design) have been adapted to slow soaring.

White-tailed eagle (*Haliaetus albicilla*): $W = 50$ N, $S = 0.72$ m^2, $b = 2.20$ m.

daring soaring exhibitions were performed by the gulls there, exactly as I had expected. But occasionally a jackdaw would join the fun, matching the various stunts of the gulls. It did not cross my mind that these jackdaws couldn't have competed with the gulls if their aerodynamic performance had been far inferior. By the time I realized my misapprehension, many years later, a friendly biologist had explained to me that jackdaws are cliffs dwellers by origin, and that the evolution of the species would have given them quite respectable soaring characteristics.

The jackdaw in Lund had a weight of 1.8 newton, a wing area of 0.059 square meter, and a wingspan of 0.60 meter. The cruising speed computed from equation 2 in chapter 1 is 9 meters per second. The flight speed for the best glide ratio in the wind tunnel was 8.5 meters per second, not too different from the number obtained with equation 2. At that speed the glide ratio was 12.5, the number that surprised me so much.

The fantastic glide ratio of jackdaws was a pleasant surprise, but what bothered me was that the migration speed of jackdaws is much higher than 8.5 meters per second. In 2001 the Swiss Ornithological Institute reported 15 meters per second (34 miles per hour), and in 2007 the Lund University group reported 12.5 meters per second (28 miles per hour). I first attempted to reconcile these numbers by extrapolating the observed gliding performance to high speeds, but that didn't help. Wings designed for 9 meters per second create far too much drag at higher speeds. What had I done wrong? I hadn't taken into account that flapping wings are like two-stroke engines. The downstroke is the power stroke; in the upstroke wings recuperate from the work just done. How to perform the downstroke without undue energy loss? Not by forcing the wings to do their job at a too high angle of attack. Not only would that bring the wings close to stalling, but it also would greatly increase the induced drag (see equation 30). It is much smarter to keep the angle of attack during the downstroke equal to the one that gives the best glide ratio. Since the downstroke has to produce a lift force equal to twice a bird's weight, and the lift is proportional to the square of the speed, the proper solution is to fly 40 percent faster. This is a drastic simplification, but it allows me to make the jump needed for a first attempt at reconciling the observed gliding performance with the observed migration speeds.

I decided to resolve the matter by making a graph (figure 18). I plotted the glide data from the wind tunnel in two ways: as drag versus speed (curve A) and as power versus speed (curve C). Curve A is equivalent to the jackdaw curve in figure 15, but upside down. In gliding, gravity provides the power needed to maintain speed. At 7.5 meters per second, the rate of descent of the jackdaw was 0.6 meter per second. The power supplied by gravity to a 1.8-newton bird then was $0.6 \times 1.8 = 1.08$ watt, as figure 17 shows. The next thing I did was to slide the drag curve 40 percent to the

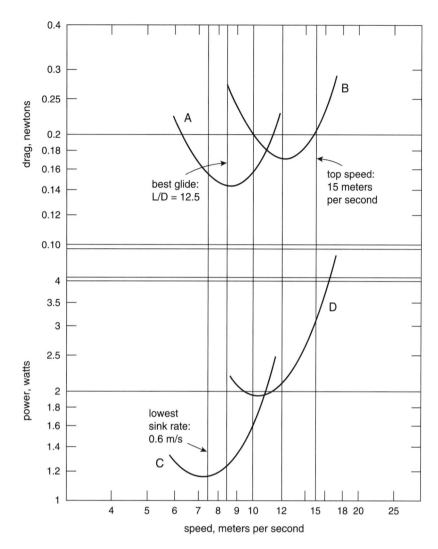

Figure 18 Observed gliding performance and estimated flapping performance of the jackdaw in the wind tunnel at Lund University. Curve A is the drag computed from the wind-tunnel data. The drag is lowest at a flight speed of 8.4 meters per second. The corresponding glide ratio is 12.5. Curve B is the estimated drag during flapping flight. It is obtained from curve A by shifting it 40 percent to the right, and giving it a 20 percent penalty to account for upstroke losses. Curve C gives the power provided by gravity in gliding flight. It is obtained from curve A by the formula $P = DV$ (power equals drag times speed). Curve D is the estimated power required for flapping flight, obtained in the same way from curve B. At a flight speed of 12 meters per second, flapping flight is most economical. The power needed at that speed is a bit more than 2 watts. The estimated maximum continuous power from the flight muscles is 3.6 watts, permitting migration speeds up to 16 meters per second.

Blue underwing (*Catocala fraxini*): $W = 0.012$ N, $S = 0.0027$ m², $b = 0.08$ m.

right, adding a 20 percent penalty to account for upstroke losses. This procedure resulted in curve B, which has its minimum at 12 meters per second, not far from the migration speed recorded by the Swedes. Minimum drag means minimum fuel costs per mile, so 12 meters per second is appropriate for migrating jackdaws not in a desperate hurry. Curve B is an estimate, but it isn't in gross disagreement with the sparse facts.

Using $P = DV$ to convert drag to power, I obtained curve D from curve B. The difference in power level between gliding and flapping is remarkable. Almost 1.2 watt is enough when gravity does the work, but 2 watts or more are needed in flapping flight. How much power can be supplied by the flight muscles? The rule of thumb is that the pectoral muscles of a bird can deliver 20 watts per kilogram of body mass. For a 180-gram bird this computes as 3.6 watts of maximum continuous power. Comparing this with curve D in figure 17, I conclude that jackdaws can manage 16 meters per second if they are behind schedule or if they run into headwinds. The speed reported by the Swiss radar crew is 15 meters per second, just a little less than the computed top speed.

I am aware the computed numbers presented are merely estimates. I would much prefer actual wind-tunnel data on flapping jackdaws, but no such data are available. The numbers given here are mutually consistent, however, and provide a plausible explanation for the difference between the best gliding speeds and the speeds chosen during migration.

Hummingbirds and Other Hoverers

Hummingbirds and many insects sip nectar from flowers while hovering in the air. Relative to forward flight, this way of living requires a lot of energy. As might be expected from equation 26, the induced power P_i becomes very large when the forward speed V becomes zero. Some calculations are in order.

For the second time we use the principle that a force equals the product of a velocity imparted by an object and a mass flux. The mass flux generated by the buzzing wings of a hummingbird is about one-fourth of dwb^2, where w is now the downward velocity in the jet of air that keeps the bird aloft. It is useful to compare this with equation 23, where the airspeed V plays the role that w plays here. The aerodynamic lift generated by the momentum transfer to the jet is

$$W = 0.25\rho w^2 b^2. \tag{33}$$

Hummingbirds and most insects do not have particularly slender wings; $A = b^2/S$ typically has a value around 6. Substituting this into equation 33, we obtain

$$W = 1.5\rho w^2 S. \tag{34}$$

Again we need equation 1, which reads

$$W = 0.3\rho V^2 S.$$

This allows us to compare the downward velocity w in the jet of air by which the hovering bird or insect keeps itself aloft with the nominal cruising speed V:

$$w = 0.45 V. \tag{35}$$

This relation explains why hummingbirds, wasps, bees, and beetles have computed cruising speeds of roughly 7 meters per second. When we substitute $V = 7$ meters per second into equation 35, we obtain

$$w = 3 \text{ meters per second.}$$

Mallard (*Anas platyrhynchos*): $W = 11$ N, $S = 0.093$ m^2, $b = 0.9$ m.

We have seen this type of expression before. The rate of descent of a gliding bird is a measure of the specific power, P/W, that it needs in horizontal flight (equation 21). For hovering birds and insects, the downward velocity of the airstream generated by the wings plays exactly the same role. Since the continuous power rating of flight muscles is about 100 watts per kilogram, and since about 30 percent of the overall weight of hovering birds and insects consists of flight muscles, the specific power output is about 30 watts per kilogram of overall weight, or 3 watts per newton. But watts per newton equal meters per second. At full power, therefore, hummingbirds and bees can generate a jet with a velocity of 3 meters per second to keep them airborne. But that is exactly what we calculated above! In other words, hummingbirds and bees are running at full power continuously. In retrospect, the airspeed of 7 meters per second at which hummingbirds, bees, wasps, and beetles are listed in figure 2 is not primarily a measure of the cruising speed they can maintain. (They have plenty of spare power with which to fly faster if they want to.) Instead, it is a measure of the strength of the jet they can generate beneath their buzzing wings (3 meters per second).

Puffin (*Fratercula arctica*): $W = 2.7$ N, $S = 0.035$ m^2, $b = 0.56$ m.

Hummingbirds have evolved to run at full power all the time because the job of transferring momentum to the surrounding air is strenuous when the forward speed is zero. The same is true for helicopters: they can relax a little only when their forward speed is high enough. In forward flight it is much easier to transfer the required momentum to the air. How does this work out for larger birds? Isn't it true, for example, that many kinds of ducks are capable of vertical takeoff and landing? The wing loading of a mallard (*Anas platyrhynchos*) is about 120 newtons per square meter. This corresponds to a cruising speed of roughly 18 meters per second. On vertical takeoff a mallard must therefore generate a downward jet with a velocity of 8 meters per second (equation 35: $0.45 \times 18 = 8$). But that requires 8 watts of takeoff power per newton of weight. That is 4 times the continuous power rating of the flight muscles. A mallard can sustain this much power for only a few seconds. After takeoff it shifts into forward flight as soon as it can.

Now we can also understand why the largest hummingbirds are much smaller than the largest birds. For hummingbirds the upper limit is about 20 grams, while the largest birds weigh approximately 10 kilograms. Hovering is an uneconomical way of life.

How much fuel does a hummingbird consume? Sugar supplies 14 kilojoules per gram, as does honey. Nectar, half water and half honey, supplies 7 kilojoules of energy per gram. With a metabolic efficiency of 25 percent, the hummingbird's net production of mechanical energy is a little less than 2 kilojoules per gram. Now we have to calculate the power requirements. The energy transferred to the downward jet of air is $W \times w$ joules per second. A 3-gram hummingbird hovering at full power, with $w = 3$ meters per second, therefore requires a mechanical energy supply of 0.09 joule per second. After all, 0.03 newton times 3 meters per second equals 0.09 watt. Since there are 3,600 seconds in an hour, the hummingbird needs a little more than 300 joules per hour. Nectar supplies 2,000 joules per gram; thus, a gram of nectar suffices for 6 hours of flying. But this implies that a 3-gram bird consumes its own weight in fuel every 18 hours! One hopes that hummingbirds are permitted some rest at night, because a full day's work in the tropical rain forest requires two-thirds of its weight in nectar. This kind of luxury is feasible only in the overwhelming extravagance of a tropical ecosystem, where flowers bloom abundantly throughout the year.

In view of the high fuel consumption of hummingbirds, it is even more amazing that some species migrate over long distances. Every autumn the ruby-throated hummingbird travels from the United States to Central America, crossing the Gulf of Mexico on its way, and every spring it retraces its route. It could not manage if it could not switch from burning sugars to burning fats on long hauls. The journey across the Gulf of Mexico takes about 30 hours (500 miles at 17 miles per hour). I can't imagine how the hummingbird could perform that feat on sugar water alone, though it is smart enough to wait for a firm tailwind before it takes off.

Gossamer Albatross: $W = 940$ N, $S = 70$ m^2, $b = 29$ m.

Among the redeeming qualities of our species is that we play. Indeed, we surround ourselves with toys, and we remain preoccupied with them throughout life. Flying toys range from paper airplanes, kites (some simple, many exotic), boomerangs, frisbees, aerobees, and spring-powered toy birds to radio-controlled model airplanes, full-scale hot-air balloons, blimps, airships, hang gliders, ultralights, and sailplanes. There are home-built racers, human-powered airplanes, and even solar-powered planes. We display almost inconceivable creativity as we tinker with our playthings. The force of imagination and the passion for experimenting propel us toward outrageous designs and technological experiments.

By the mid 1970s, sailplanes had achieved a finesse of 40 and a rate of descent of 60 centimeters per second (120 feet per minute). One would think that this would satisfy even the most fanatical glider pilots. (A lot of work was needed to maintain this kind of performance. Every weekend started with hours of scrubbing and polishing, and after every flight the dead bugs had to be removed from the leading edge of the wings. To keep the finesse from dropping to 20, the wings had to be perfectly smooth.) Designers didn't see much scope for progress: more sophisticated airfoil designs would require expensive wind-tunnel tests and computer simulations, and longer wingspans were not possible with the structural materials then available.

The wind-tunnel problem was not as difficult as it seemed. The smart thing to do when you need sophisticated equipment is to get in touch with professionals. Before you know it, they are as enthusiastic as you are and will use their spare time to do the necessary

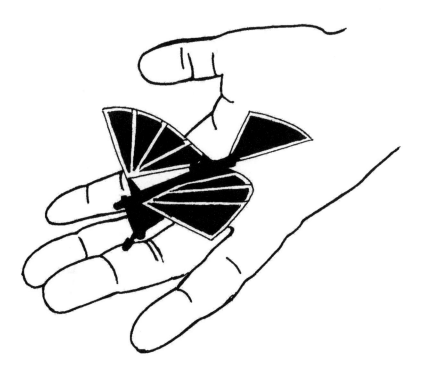

A toy ornithopter.

research. The Schleicher ASW-22B open-class competition sail-plane, for example, achieves a finesse of 60 in part because of a wingtip design that was lovingly perfected in Delft, in the same aerospace engineering department where I studied many years ago. It must have cost untold hours to wring that last bit of progress out of a mature technology.

The great breakthrough in sailplane construction came around 1980, when various new materials arrived on the scene: carbon and aramid fibers, expanding foam fillers that made "sandwich" construction possible, and adhesives that could withstand structural stress. It was some time before appropriate assembly methods were developed, but after that the designers had a field day. In airplane technology, where every ounce of superfluous weight must be avoided, such opportunities are exploited. Boeing's 747-400 saves about 7,000 pounds through the use of advanced materials and construction techniques. If it would use the saved weight to carry an additional pallet of fresh flowers across the ocean, it would earn almost $10,000 extra. Toymakers are just as quick to take advantage of space-age technology. The frame of the Revolution kite is made of "100% aerospace graphite," most kites nowadays have Dacron sails and Kevlar or Dyneema strings, the wing spars of human-powered airplanes are made from aramid fibers, and nearly all sailplanes are now built with vacuum-formed composite construction techniques, which combine low weight with great rigidity. Through such advances, today's best competition sailplanes, with a wingspan of 25 meters, achieve a finesse of 60 and a rate of descent of 40 centimeters per second (80 feet per minute). In finesse they beat albatrosses, the best nature has to offer, by a factor of 3, and in rate of descent they outperform their nearest avian competitors, swifts and swallows, by a factor of 2, even though the birds have a substantially lower weight, wing loading, and cruising speed. In fact, open-class competition gliders perform so well that they carry 200 liters of water as ballast. Aeronautical engineers are always keen to save weight, yet here are competition sailplanes taking ballast along! Why? As we saw in chapter 4, the distance covered in gliding is determined by the finesse. In turn, the finesse is determined by streamline shape, surface smoothness, and wing aspect ratio, *not* by wing loading. As the weight of an airplane increases, its speed must increase, but its finesse remains the same. Therefore, if you are in a hurry or if you want to cut your losses in a headwind, you are better off if you are overweight: that increases your cruising speed. The only sacrifice you make is that your rate of descent increases somewhat, but it was extremely low anyway. So, contrary to all aeronautical principles, you make your sailplane heavier than is strictly necessary. And in order to have the best of both worlds, you arrange it so that you can dump your ballast when the updrafts are weaker than anticipated. Water is

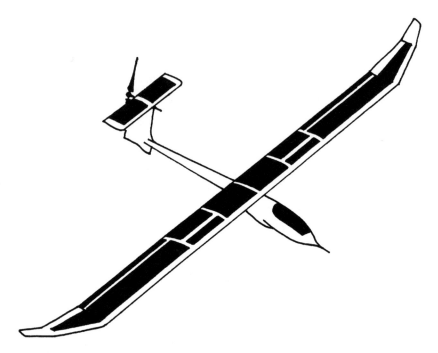

Icaré 2, a solar-powered airplane: $W = 3{,}500$ N, $S = 21$ m^2, $b = 25$ m.

useful for this purpose, since at worst it might create an unexpected shower for some innocent bystander.

A wide range of flying playthings is available. At one end we find indoor flying models, which are designed for extremely low speed. Even with a rather modest finesse ($F = 8$ at best), such a model achieves a very low rate of descent. Its wing loading is only 0.1 newton per square meter, one-tenth that of a small butterfly. The largest of these models have a 90-centimeter wingspan (3 feet), a wing area of 1,000 square centimeters, and a weight of 2 grams (less than a sugar cube). Half of that weight is accounted for by a tightly wound rubber band driving a very large and slow propeller. In a sports hall, such a model achieves a flight duration of 45 minutes, traveling only half a meter per second. Once the propeller stops turning, the model loses altitude at a rate of 5 centimeters per

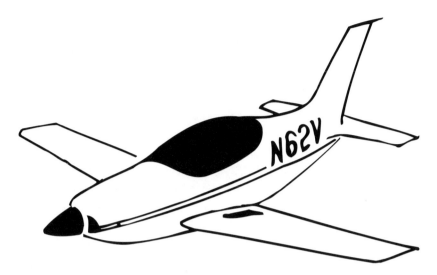

Questair Venture: $W = 8,000$ N, $S = 6.76$ m², $b = 8.40$ m, $P = 224$ kW (300 hp).

second (10 feet per minute). Its rate of descent is one-sixth that of a cabbage white. At the other end of the scale are airplanes specifically designed for racing. The Questair Venture, with a 224-kilowatt (300-horsepower) Porsche engine, reaches a top speed of 463 kilometers (290 miles) per hour on wings with a surface area of less than 7 square meters. The engine alone accounts for 30 percent of the takeoff weight. When you hit the throttle for the first time in this machine, you had better take care; it is probably just as temperamental as the late-model Hurricanes and Spitfires of World War II.

Whereas the top speed of a 300-horsepower sports car is only 160 miles per hour, a sports plane with the same engine goes nearly twice as fast. If you really get a kick from speed, you would do better to take up flying. At least you won't be a menace to others on the freeways.

You don't really need several hundred horsepower to have plenty of weekend flying fun; 30 will suffice. You can see why at any airstrip where ultralights are flown. You will find them there in many shapes and sizes—ragtag contraptions held together by steel cables and covered with spinnaker nylon. They look like hang gliders with tricycle gear and lawnmower engines. If you want to fly a hang glider, you must first find a mountain slope and wait for sufficiently strong winds, but with an ultralight you can come and go as you please, even on the plains and in quiet weather. If you don't mind spending the additional effort, you can even attach flotation gear. That gives you the opportunity to visit quiet lakes in the countryside.

Suppose you want to design your own ultralight. Allowing 70 kilograms for your own weight, 40 kilograms for the wings, 30 kilograms for the engine and the propeller, and several tens of kilograms for wires, cables, piping, frame, wheels, and a small fuel tank, you can estimate the total flying weight as 200 kilograms (440 pounds). Suppose you wish to cruise at 60 kilometers per hour (almost 17 meters per second). The first step in the design process is to consult equation 2, which shows that a wing loading of 106 newtons per square meter is required. From this number and the 2,000-newton overall weight, it is a small matter to calculate that the wing area must be 19 square meters (200 square feet). Now look at figure 15. A finesse of 8 would seem reasonably conservative; you can't pretend to aim for sophisticated streamlining. With a cruising speed just under 17 meters per second and a finesse of 8, the rate of descent of your ultralight will be a little over 2 meters per second. But the rate of descent tells you how much power you need to keep your weight airborne: $w = P/W$, as equation 21 shows. Since $W = 2,000$ newtons and $w = 2$ meters per second, this puts the power required at 4,000 watts. But that is not enough. If you want to be able to climb at a rate of 3 meters per second (600 feet per minute), you will need an additional 6,000 watts, for a total of 10,000 watts. Ten kilowatts, or 14 horsepower, doesn't seem much. But you haven't yet factored in that you are stuck with a somewhat small and fast propeller. In this application a fairly large and slow propeller would be best, but you must make do with a small propeller mounted directly to the crankshaft of a

Figure 19 Two amphibious ultralights.

fast-running engine. At best you can hope for a propeller efficiency of 50 percent, and this forces you to select a 20-kilowatt (27-horsepower) engine. Now the time has come to sit down at a drawing board and work out the details of your design, making allowances for the unforeseen so that, for example, a wing design that turns out to be slightly heavier than expected will not be disastrous.

For just about anything that flies it is a good idea to maximize the finesse, given all the other design constraints. A kite, however, doesn't really benefit from a finesse higher than 2. An aerodynamically advanced kite with slender wings will float almost straight above your head. But a kite is not stable in that position. It will behave like an errant sailplane, dangling a slack string behind. Since the string has fallen slack, a brisk tug won't help; the kite will float around until it begins to drift sideways. By the time it draws its string taut, the sideways dive of the beautiful design you worked so hard on over the weekend will have become uncontrollable.

A two-string kite can be maneuvered out of danger, and Dacron cloth and carbon-fiber spars can stand a lot more abuse than the Chinese paper and bamboo spars of earlier days. Still, kites perform best when they draw their strings taut. This requires stalled airflow over the wings—an aerodynamic condition that birds and aeronautical engineers will do anything to avoid.

An aeronautical engineer would also hesitate before selecting a steam engine to power a plane. It is far too heavy, and its thermal efficiency is hopelessly poor. Yet the first powered model airplane in the world was driven by steam. In 1896 the American aviation pioneer Samuel Pierpont Langley flew a powered model airplane across the Potomac River near Washington, a distance of more than two-thirds of a mile. The craft had tandem wings spanning 12 feet and weighed 30 pounds. With its boiler, the steam engine weighed 7 pounds. The power delivered to the propeller was probably only 300 watts, one-tenth the power output of a combustion engine of the same weight, but the event was a significant step forward in the history of aviation. Seven years later, on December 17, 1903, the Wright brothers made their historic flight at Kitty Hawk.

A flock of geese flying in formation with an ultralight airplane, taking advantage of the updrafts caused by the tip vortices.

Trials with a Paper Airplane

A book on flight would not be complete without a few pages on paper airplanes. Who hasn't played with a paper plane at one time or another? Without getting distracted by intricate folding techniques, you can easily make a paper plane that concentrates on the art of flying. All you need is a 4 × 6-inch index card and a large paper clip (figure 20). (A smaller, thinner piece of paper and an ordinary paper clip are suitable, too; there is room enough for experimentation here.) You will see why you need the paper clip as soon as you attempt to launch your model plane. You do that by giving it a push forward, at a speed comparable to that of walking. But it doesn't want to fly; it starts to tumble backward. The reason is that the center of gravity of the index card lies in the middle of the paper, while the aerodynamic forces are located in front of the center of gravity. The problem is solved by shifting the center of gravity of the index card forward—hence the paper clip. The plane's center of gravity should be located at roughly 30 percent of the distance between the leading and trailing edges of the wing.

Don't make any fold or crease in the index card yet. Position the paper clip carefully, making sure that it sits exactly on the centerline, and resume your flight trials. You will discover soon enough that a paper airplane is extremely sensitive to the exact location of the center of gravity. If you move the paper clip just a little too far forward, the plane dives into the ground; if you slide it back even a fraction of an inch, the plane can't seem to settle down to a smooth and steady glide. As the nose moves up, the plane loses speed until it stalls, its nose drops, and its speed increases. But then the nose moves up again, and the process repeats itself. If the center of gravity is a little too far to the rear, the paper airplane behaves exactly like the juvenile herring gull mentioned in chapter 3.

Your airplane will be in proper trim when the center of gravity is in the right place. It will then glide smoothly. Knowing that a very high finesse cannot be expected from a simple piece of paper, you may note with satisfaction that your plane glides 4 feet for each foot of altitude lost. $F = 4$: not bad for a toy. But not all is well yet.

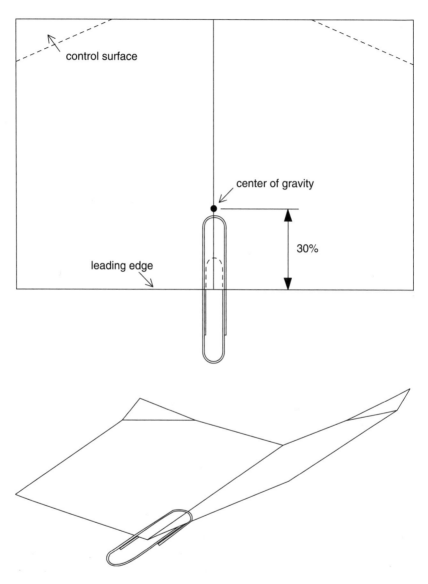

Figure 20 A simple paper airplane. The center of gravity should be located at about 30 percent of the chord (the distance between the leading edge and the trailing edge). The control surfaces in back can be used both as elevators and as ailerons. Once you have adjusted the position of the paper clip for optimum performance, fix it with a piece of tape.

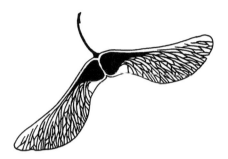

A Norway maple samara (*Acer platanoides*): $W = 0.002$ N, $S = 0.0015$ m^2, $b = 0.10$ m.

From time to time your plane starts sliding sideways, slowly at first but gradually faster. Experimenting in a gym or in the stairwell of a large office building, you will find that your plane tends to accelerate into a high-speed "spiral dive." It has a disappointing instability, and it cannot maintain a stable course. But exactly what is going wrong? Since there is still no fold in the index card, your airplane is just a flat piece of paper; it can slide sideways through the air without meeting any resistance. As figure 20 shows, you can correct this by simply making a crease along the centerline. The wingtips are slightly higher now; if the plane tends to slide to the left, the left wing will be pushed up some and the right wing pulled down. The plane rolls to the right, starting a right turn to correct for the sideslip to the left, exactly as a bicyclist counters a gust of crosswind. This is the remedy for the spiral dive. Your plane is now directionally stable.

Once you have caught the spirit of experimentation, you will do well to fix the paper clip with tape. It is also a good idea to fold the trailing-edge corners of the wings up a little so they will serve as control surfaces and make the plane fly slower. Don't overdo it, though; the plane is very sensitive to changes in control-surface angle. With some dexterity and a bit of patience you can make your plane fly near the stalling speed. The control surfaces will also come in handy if the paper clip shifts a little after a crash landing

A flock of whooping cranes following an ultralight at a comfortable flight speed of 40 miles per hour.

into the furniture. You need not put it back to the original position, since changes in control-surface angle can easily compensate for minor shifts in the center of gravity. An airliner corrects for differences in its load distribution in the same way: if the center of gravity moves forward, the elevators (the movable parts of the tail) turn up a few degrees to keep the nose of the airplane up. That is how, notwithstanding variations in its load distribution, an airliner can keep its balance. But there is not much room for error, for the center of gravity must stay within narrow limits.

The control surfaces of your paper airplane can also be used to make it turn. If you want it to turn left, turn the left corner up a little; if you want it to turn right, do the same on the right side. Now you are using the control surfaces as ailerons. Small aileron deflections are needed also to make sure that your plane keeps flying straight if a minor mishap should make it somewhat asymmetrical.

There is one more kind of instability that you may run into: if your paper plane finds it difficult to fly at constant speed, it suffers from what aeronautical engineers call the "phugoid." This is best corrected by shifting the center of gravity forward and turning the control surfaces further upward. In any event, you should avoid shifting the center of gravity so far back that the control surfaces must be bent down to maintain trim. That is asking for disaster.

Pedal Power

People have always dreamed of flying under their own muscle power. In Greek mythology there is the famous story of Daedalus, an inventor in the court of King Minos. Because he had assisted Ariadne in arranging for Theseus to escape, Daedalus was imprisoned in a labyrinth of his own design. But he and his son Icarus managed to flee from Crete by constructing wings made of goose feathers and beeswax. Icarus, the story tells, flew too close to the sun, and the wax melted. He crashed and drowned. Daedalus, on the other hand, managed to reach the continent safely. Many centuries later, Leonardo da Vinci made sketches of helicopters and human-powered airplanes. The flapping wings he tried to design were not a very sound idea: our leg muscles are much more powerful than our arm muscles, not to mention the construction problems encountered when designing oversize flapping wings. But the dreaming continued; around 1990 Otto Lilienthal conducted his famous experiments with hang gliders.

The development of human-powered airplanes began in earnest after World War I. In the years before the war, the aforementioned Ludwig Prandtl of Göttingen, Germany, one of the founders of modern aerodynamics, had systematized the basic principles of flight. The spectacular progress achieved by Prandtl and his colleagues inspired several German universities to include aeronautical engineering in their curricula, and by 1914 aeronautics had become a popular discipline.

World War I brought the first large-scale use of military airplanes. When the war was over, Germany was prohibited by the Treaty of Versailles from engaging in the design or production of war machinery. As a result, German aeronautical engineers were limited to designing unpowered airplanes. Then, as now, designing an actual airplane formed an integral part of the curriculum for senior-year students. Because fighters and bombers were out of the question, German students focused on gliders. To this day, European soaring competitions are dominated by teams from universities in Berlin, Darmstadt, Braunschweig, Karlsruhe, and Stuttgart. Each year, the professors dream up new senior-year design projects in the hope of perfecting their super-sophisticated toys.

In the period between World War I and World War II, the efforts of German aeronautical engineers resulted in well-engineered gliders with creditable performance for their time. The Germans also attempted to design human-powered airplanes, but these were failures. After World War II there was a revival of interest in human-powered airplanes in England. This fire was kindled by a prize of £5,000 offered in 1959 by the industrial tycoon Henry Kremer. It would be awarded to the first pilot who would complete a figure-eight pattern between two posts half a mile apart. A group at Southampton University tried first, but failed. Then the Hatfield Man-Powered Aircraft Club took up the challenge. They built their *Puffin* (figure 21) within two years. It flew it in the autumn of 1961. *Puffin* was a high-tech plane for its time, but the design choices were all wrong. The design team failed to do the simple calculations that would have told them that an extremely low airspeed was required for a decent chance of success. Instead, they believed they had good reasons to chose a wing area of 31 square meters. With a total weight of 260 pounds (1,200 newtons), *Puffin*'s airspeed computed at 10 meters per second (22 miles per hour), far too fast for human athletes. Speeds like that cannot even be maintained for any length of time by amateurs on racing bicycles. One way to confirm these numbers is to start with an estimate for the finesse of *Puffin*. I put that at 30, meaning that the aerodynamic drag is $\frac{1}{30}$ of 1,200 newtons—that is, 40 newtons. Drag times speed equals power, so *Puffin* required 400 watts to stay in the air. Lance Armstrong could have managed, but most bicycle racers still can't,

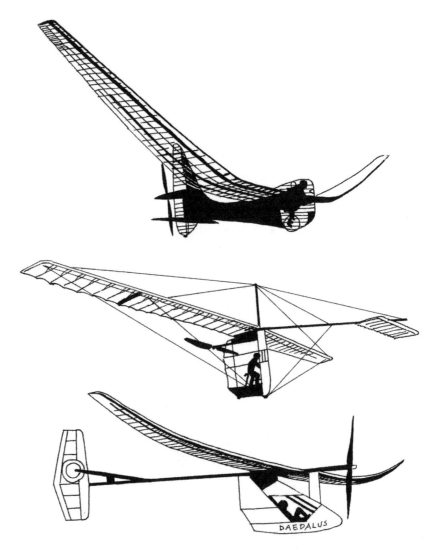

Figure 21 Three human-powered airplanes. From top: the 1961 *Puffin*, the 1979 *Gossamer Condor*, and the 1988 *Daedalus*.

not even for a few minutes. The longest hop *Puffin* managed was 900 meters, a little over half a mile.

The years passed, but no contender was successful. Kremer gradually increased the stakes, and by 1975 the prize had risen to £50,000. That was enough to stimulate some serious thinking on the part of Paul McCready, an aeronautical engineer and president of the California environmental consulting firm Aerovironment. He started by considering available muscle power. Provided that the effort lasts no longer than several minutes, a well-trained athlete can attain a power output of 250 watts. McCready also realized that the weight of the airplane must be kept low. A heavy plane must fly fast, thus requiring too much power. McCready decided that the takeoff weight could not exceed 100 kilograms. With a 65-kilogram bicycle racer on the pedals, this left only 35 kilograms for the airplane. And since one can't expect aerodynamic perfection from a lightweight contraption made of corrugated cardboard, piano wire, and Saran Wrap, one can't achieve a very high finesse. McCready chose $F = 20$, not the $F = 40$ that had become common for sailplanes.

If the total weight is 100 kilograms and the finesse is 20, then the drag is 5 kilograms, or 50 newtons. It was at this point that McCready made the crucial computation: if you have 250 watts to offer and you have to overcome a resistance of 50 newtons, at what speed can you travel? Since a watt is a newton-meter per second, 250 watts will give you a maximum speed of 5 meters per second against a 50-newton drag. (See chapters 2 and 4.) Five meters per second, or a little over 10 miles per hour, is typical of a teenager on a "granny bike." To keep 100 kilograms airborne at a speed of 5 meters per second requires extremely large wings. The rule of thumb from chapter 1 is

$$W/S = 0.38V^2.$$

With $W = 1,000$ newtons and $V = 5$ meters per second, the wing area S would have to be more than 100 square meters (1,100 square feet)—roughly the floor space of a small two-bedroom apartment.

McCready decided that he could achieve the same result with 70 square meters of wing area if the plane were to fly at a speed very close to stalling.

If a plane is to achieve a finesse of 20, its wings must be very slender. According to the data in chapter 4, the aspect ratio must be at least 12 to yield the desired result. On this basis, McCready calculated that his plane would need a wingspan of 30 meters (100 feet). That's the height of a ten-story apartment building; that's twice the wingspan of a standard-class glider; that's almost as long as two tractor-trailers; that's the full length of a high school gymnasium! Just think of the design job: 100 feet of wing that must not weigh more than 25 kilograms, because the last 10 kilograms must be reserved for a bicycle frame, pedals, gears, chains, and a propeller. McCready's team succeeded nevertheless, and in the early morning of August 23, 1977, the Kremer Prize was won.

With a multitude of problems to overcome, McCready often came close to abandoning the project. Accidental gusts caused several crashes; after a while, all flight trials were conducted before dawn. Fortunately, the primitive construction methods used by his team allowed for quick repairs; McCready said he probably wouldn't have persevered if repairs had taken more time. He was also lucky to have several friends who worked in the Graduate Aeronautical Laboratories of the California Institute of Technology and at the University of Southern California. In the early stages of the project, McCready could not design a suitable propeller. With only one-third horsepower available, he could not afford any energy losses. Professor Peter Lissaman helped him out by writing a sophisticated computer program to optimize the propeller design. A few unauthorized weekend runs on USC's supercomputer quickly solved the problem.

The board of directors of the giant chemical company DuPont, prepared as ever to support intelligent dreamers, offered to sponsor McCready's project and to supply advanced materials, including Mylar, Kevlar, and carbon fibers. Meanwhile, Peter Lissaman made additional computer calculations, and other associates concocted construction methods that reduced the amount of piano wire and thus the aerodynamic drag. The finesse inched its way upward to 25, while the empty weight of the plane came down at least 6 kilograms. With $W = 940$ newtons and $F = 25$, the drag was not 50 newtons but 38; at a speed of 5 meters per second the power required was only 190 watts. A professional bicycle racer in good condition could keep that up for several hours. Long-distance trips were coming within reach.

The McCready team was responding to Henry Kremer's latest challenge. Kremer had announced a prize of £100,000 for the first human-powered flight across the English Channel from Dover to Calais. That prize was won by Bryan Allen, the same cyclist who had captured the first one two years earlier. On June 12, 1979, he pedaled the *Gossamer Albatross* from England to France. Allen counted on having to pedal for just under 2 hours, but near Cap Griz-Nez he encountered unexpected headwinds. Because of the wind, it took him 2 hours and 45 minutes to reach the other side.

Kremer just couldn't stop teasing the human-powered-airplane crowd. He thought McCready's planes were far too slow, which made them sensitive to turbulence, gusts, and headwinds. What's the use of a plane that can be flown only at dawn? So he offered yet another prize: £20,000 for the first human-powered airplane to complete a one-mile triangular course within 3 minutes. This would require a speed of more than 20 miles per hour. Engineers and scientists at the Massachusetts Institute of Technology used their ingenuity to achieve a still greater finesse than the *Gossamer Albatross*. Twenty miles per hour is about 10 meters per second; at that speed, a finesse of 33 is needed to bring the drag of a 100-kilogram plane down to 30 newtons. The power required then is 300 watts, a rate a professional bicyclist can maintain for 10 minutes at best. In May 1984, the MIT team's plane, named *Monarch*, won the prize.

The dream of Daedalus came true on April 23, 1988, when the Greek cyclist Kannellos Kannellopoulos managed to fly from Crete to the island of Santorini, 120 kilometers away. The plane, named

Daedalus, had a wingspan of 34 meters, a wing area of 31 square meters, and a finesse of 38. It had been designed and built by same MIT team (led by Mark Drela and John Langford) that had built *Monarch*. United Technologies had sponsored the project with half a million dollars. With a cruising speed of 7 meters per second, 210 watts was sufficient—too much to demand of an amateur, but within reach of the professional Kannellopoulos. It had taken a team of physiologists, ergonomists, and other experts 4 years to select and train the winning pilot. Several others had been eliminated from the competition because of inefficient metabolism or poor muscle discipline. Great strength in itself is not necessarily an asset when it comes to flying.

Solar Power

Once Paul McCready had won the two Kremer Prizes, he took to designing solar-powered airplanes. His *Gossamer Penguin*, piloted by his teenage son, managed a few brief flights on the California desert, and on July 7, 1981 his *Solar Challenger* flew from Paris to London with a human pilot in the cockpit. With financial support from NASA, McCready also designed and built the unmanned *Pathfinder*, a 250-kilogram solar-powered flying wing with a 30-meter wingspan and six propellers, which first flew in 1993. It was followed by a string of successors and competitors, designed to fly day and night far above airline traffic. The proponents were dreaming of "eternal flight," with unmanned solar planes recharging their batteries in daylight and continuing on battery power at night.

The ultimate challenge for solar-powered flight came when, in 1994, the German city of Ulm offered a prize of 100,000 Deutsche Marks for a piloted airplane that would have to be capable of level flight at half the 1,000 watts per square meter of solar power available on a sunny day and would have to withstand diving speeds up to 70 miles per hour. The toughest requirement was that the plane would have to take off and climb 2 meters per second (400 feet per minute) until it reached an altitude of 1,500 feet. With a pilot, such a plane would easily weigh 350 kilograms, as much as a standard-class sailplane. Lifting this weight (which amounts to 3,500 newtons) at a rate of 2 meters per second requires a power of $2 \times 3,500 = 7,000$ watts. That is about 10 horsepower, far too much to be supplied by solar cells. Obviously a pack of batteries was needed. But those would add to the weight, making the design puzzle harder yet.

Many groups decided to participate in the Ulm competition, but only four showed up at the fly-off in July 1996, and only one aircraft actually managed to fly. It was *Icaré 2*, designed by a team from the University of Stuttgart. It had a wingspan of 25 meters, a wing area of 25 square meters, and a total weight, including the pilot, of 3,600 newtons (almost 800 pounds). At the design speed of 17 meters per second (28 miles per hour) it had a glide ratio of 40, so the drag was 90 newtons. Therefore, the power required was $90 \times 17 = 1,530$ watts (roughly 2 horsepower). Twenty-one square meters of solar cells, with an efficiency of 17 percent, delivered about 1,800 watts at the specified solar energy input. After deductions for transmission and propeller losses, only 1,600 watts of power were available. But think of the opportunities! Roll your solar-powered plane out on a sunny morning and relax on the terrace of the airport's diner. While you're having a cup of coffee, your plane, like a giant dragonfly, is basking in the sunlight and recharging its batteries. Then take off on battery power and climb to 1,500 feet. On a sunny day, the power output of the solar cells is twice as much as was specified for the Ulm competition: about 3,000 watts (4 horsepower). So you can climb higher if you want, or you can make a high-speed crossing to the next thermal, where you can soar and recharge the batteries at the same time.

Solar power has inspired toymakers, too. Many kinds of miniature solar-powered planes have appeared on the market in recent years. Because the energy-conversion efficiency of commercially available solar cells is low, these toys have to fly slowly. They typically have one-fourth the wing loading and one-half the flight speed of birds of the same weight (figure 2). The smallest—weighing only 4 ounces, a little more than a newton—fly no faster than the *Gossamer Condor*: about 5 meters per second (11 miles per hour). A large cluster of solar toy planes centers around a weight of 10 newtons (somewhat more than 2 pounds), with a wing loading of 25 newton per square meter, a wingspan of 5 feet, and a speed of 8 meters per second (18 miles per hour).

Failure and Success

The story of engineering progress is not complete if the endless series of mishaps that pave the way toward success is ignored.

In the late 1990s a group of Boeing engineers attempted to design and produce a human-powered airplane that would be able to outperform *Daedalus*. Their *Raven*, constructed from black carbon-fiber composites, was intended to achieve a flight duration of 5 hours, an hour more than *Daedalus*. But construction problems and insufficient funding grounded the project.

All too often, airplane designers are not conservative enough. They risk running into trouble when they stretch the limits of technology. *Raven* is a case in point; so is Beechcraft's *Starship*. The *Starship* had a "canard" wing in the rear, and control surfaces up front. That caused unfamiliar stability and control problems. The futuristic-looking *Starship* did not sell well. It is worth contemplating why all long-distance airliners still copy the engine mounts of the Boeing B-47 bomber, which first flew in 1947. Engines hung in pods below the wings can be shed when they shake apart, and an engine fire does not threaten the passenger cabin. All other design options have disappeared. Another chronic design problem is that new jet fighters are loaded with so many options that they become far too heavy. I attribute the worldwide success of the General Dynamics F-16 to the company's decision to design a lightweight fighter. That decision reversed the trend toward ever-heavier weapons systems. A classical story about designers taking irresponsible risks is told in Nevil Shute's 1948 book *No Highway*, in which the fictional Reindeer airliner suffers fatal fatigue cracks.

In 1982, Henry Petroski, the author of several perceptive books on the evolution of engineering, published *To Engineer Is Human: The Role of Failure in Successful Design*. His most recent book, dating from 2006, is called *Success through Failure: The Paradox of Design*. Petroski attributes the collapse of the Tacoma Narrows bridge in 1940 to its stubborn chief designer, who underestimated the instabilities of a narrow two-lane road deck. The Golden Gate bridge, with its four-lane deck, is a lot safer, though it came close to collapsing when a quarter of a million people converged on it to celebrate the bicentennial in 1976. Petroski's message is clear: success tends to make us overconfident; failures force us to rethink our assumptions. In the words of the philosopher Karl Popper, "we can learn from our mistakes." If you don't, you may run into the narrow margins of success.

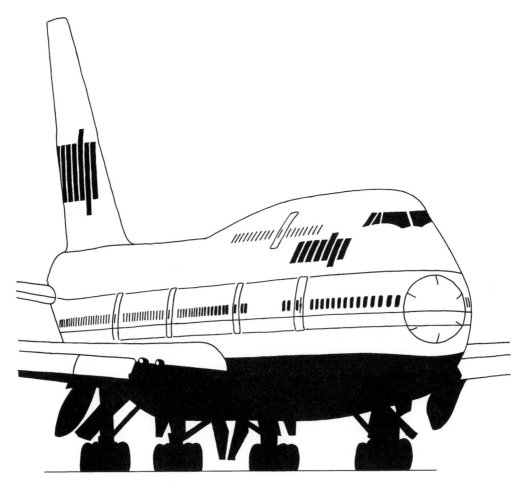

Boeing 747-400: $W = 3.95 \times 10^6$ N, $S = 530$ m^2, $b = 65$ m.

On November 23, 1991, en route to Washington, I was standing in the cockpit of a British Airways Concorde, chatting with the flight engineer. We were flying at 58,000 feet at twice the speed of sound (22 miles a minute). The sun had just risen above the *western* horizon. I scanned the fuel gauges but couldn't find what I was looking for.

"What is your fuel flow?" I asked.

"Twenty tons per hour," the engineer replied.

"That's twice the fuel flow of a 747," I said.

"Yes, but we're going twice as fast."

A Concorde needed a lot of fuel to carry 100 passengers across the ocean. In fact, 3 hours in the air required 77 tons of kerosene. Allowing for a few empty seats, this comes to 1,000 liters per person. A 747 also consumes about 70 tons of fuel between Heathrow and Washington-Dulles, but it carries 350 people and 30 tons of freight along on the trip. The Concorde could not afford to carry any freight and its range was only 7,000 kilometers. Its tanks had to be filled to their full capacity before it could take British jet-setters to Barbados.

My relationship with the Concorde is an uneasy one. It dates from 1964, well before the plane's first flight. Preliminary designs for the Concorde and its American and Russian competitors were on the drawing boards at the time. I argued at my doctoral thesis defense (much to the chagrin of the professors on the examination committee) that supersonic airliners would be a step backward in the history of aviation.

The finesse (glide ratio, in conventional parlance) of aircraft designed to penetrate the sound barrier is hopelessly low. A

Boeing 777 has a finesse of almost 20, but the Concorde barely reaches 6. In 1964 there were hopes that this disadvantage could be compensated by the high thermal efficiency of supersonic jet engines, but the high-bypass-ratio engines of present-day subsonic jetliners, twice as efficient as their early counterparts, easily outperform supersonic jet engines. The Concorde looks much more elegant than any other airliner, but its performance is not elegant at all.

An airplane flying faster than the speed of sound creates shock waves in the air, much like the bow and stern waves of a tugboat crossing a harbor at speed. This is what causes "sonic booms." Creating these waves takes a lot of energy. Because a plane in supersonic flight can't avoid making shock waves, the problem of declining finesse is insoluble. Although Concorde passengers didn't notice anything as their plane penetrated the sound barrier, the economic barrier was real enough. If you want to exceed Mach 1, it will cost you 3 times as much as staying below the speed of sound. For the aircraft industry, supersonic flight was indeed a step in the wrong direction. Time and again, before aeronautical engineers started dabbling with supersonic flight, they had managed to reach higher speeds at lower costs. The Concorde broke that trend.

Farewell Concorde; Back to Common Sense

In retrospect, the Concorde was an ill-advised prestige project of the British and French governments. Commercially, it was a failure from the very start. The price of Concorde tickets was outrageous. (I profited from an off-season discount rate, with a return trip by 747.) Only 20 Concordes were built. Since all of them were in fact prototypes, they never outgrew their teething problems, and they required excessive maintenance. The curtain fell after the fiery crash (caused by tire debris on the runway that ruptured a fuel tank) of an Air France Concorde in a Paris suburb on July 25, 2000. British Airways quit flying Concordes in October of 2003.

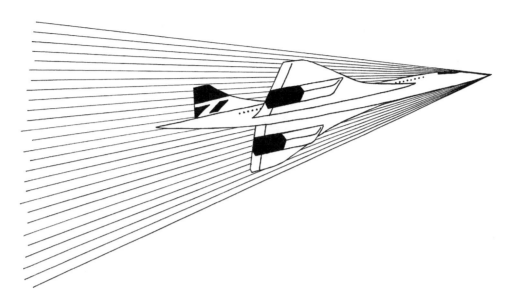

Concorde ($W = 1.8 \times 10^6$ N, $S = 358$ m^2, $b = 25.6$ m). Lines represent shock waves.

The moral of the Concorde story, as I see it, is that dreams of progress and glory are dangerous. Megalomania was the force behind several aviation enterprises that failed. Howard Hughes's "Spruce Goose," which in 1947 made a brief flight over San Diego Harbor, is one example. With a takeoff weight of more than 150 tons, a wing area of 1,060 square meters (twice that of a 747), and a wingspan of 98 meters, it was powered by eight 3,000-hp engines. Other airplanes that did not fulfill the expectations of their designers are the Lockheed Constitution, the Convair B-36 bomber, the Bristol Brabazon, and various earlier French, Italian, and German planes. If you want to know more, Google the Caproni 60, which had *nine* wings, or the Dornier X, which sported *twelve* engines. Seen from an evolutionary perspective, all these planes were misfits.

Progress in aviation came in an entirely different way. In 1965, Juan Trippe, the strong-willed president of Pan American Airways, negotiated with Boeing for a wide-bodied jet that could easily be converted to a cargo plane after a takeover of the passenger market by supersonic airliners (which was then thought likely). Trippe was not only badly mistaken about the prospect of supersonic flight; he also misread the future of low-cost, long-distance mass transportation. Pan American had quite a reputation for luxurious travel among movie stars, statesmen, royalty, and business executives; cheap air travel for the millions was not Trippe's ultimate dream. Nevertheless, his insistence forced Boeing to design the airliner that changed the world of long-distance air travel: the 747.

Boxed in by Constraints

An airliner must fly as fast as is possible without a major sacrifice in finesse. The higher its speed, the smaller the capital expenses per ton-mile or passenger-mile of travel. If depreciation alone is going to cost you several million dollars a year, you can't afford to drag your feet. Incidentally, this is the argument that kills recurrent dreams of travel by airship. One hundred miles per hour just won't do.

A second constraint is that an airliner has to fly slower than the speed of sound or else it will suffer from the drastic drop in finesse that occurs at supersonic speed. If supersonic flight didn't require so much fuel, it would be great. However, it doesn't work out that way. Mach 0.9 is an absolute maximum. These two constraints allow no latitude: a cruising speed of about Mach 0.85 is both the minimum and the maximum for long-distance airliners. But there is a bonus here: jet engines are more efficient when they fly faster. Speed acts like a turbo compressor, so at Mach 0.85 jet engines enjoy an appreciable amount of turbo boost. It comes as no surprise that the speed of jetliners hasn't changed since the first of them, the Boeing 707 and the Douglas DC8, appeared on the market in the late 1950s. In fact, the design speed of airliners has *decreased* somewhat since the 747 arrived on the market. The 747 was designed for 1,000 kilometers per hour (620 miles per hour), but its cruising speed was lowered to 900 kilometers per hour (560

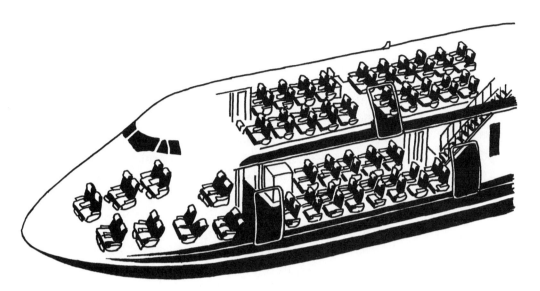

Interior arrangement of a Boeing 747-400.

miles per hour) to reduce fuel consumption. Later jetliners were designed for 900 kilometers per hour from the start. How can you confirm this? The wings of modern Airbuses and Boeings are a bit less raked than those of a 747. The "sweep-back angle" of their wings is about 30°, not 35° as on the 707 and the 747.

A third constraint arises because jet engines perform best in very cold air. The efficiency of jet engines improves as the difference between the intake temperature and the combustion temperature increases. The efficiency of converting heat into useful work depends on temperature differences, as was discovered by the French engineer Sadi Carnot (1796–1832). Cold intake air and extremely high turbine temperatures (up to 2,500°F nowadays) make for superior engine efficiency. This has far-reaching consequences for jetliners. The coldest air is found in the lower stratosphere, above 10 kilometers (33,000 feet). The temperature there is about −55°C. Long-distance planes must fly high in order to travel far, but again there is a bonus: clouds and thunderstorms are extremely rare in the stratosphere, so flight schedules can disregard meteorological conditions. In the stratosphere, airplanes fly "above the weather." This is advantageous for passengers too: above the weather only occasional turbulence is encountered.

Flying high has yet another advantage. At high altitudes, the air is much less dense than at sea level. In order to remain airborne at 33,000 feet, an airplane needs large wings. Near the ground, large wings permit lower speeds: the sea-level cruising speed is about one-half the cruising speed at altitude. If the wings are fitted with extensive flaps and slats, the takeoff and landing speeds are lower yet. This helps to limit the required runway length.

A fourth constraint is that an airplane shouldn't fly any higher than is necessary. Rarefied air requires oversize wings. Moreover, jet engines operating in rarefied air suffer respiration problems. If you insist on flying too high, your plane needs not only oversize wings but also oversize engines.

Other design factors may interfere with the fourth constraint. The Concorde, for example, had oversize wings because it had to take off from and land on the same runways as subsonic jetliners. In order to minimize the disadvantages of those wings, the Concorde had to fly high in the stratosphere.

Together, the third and fourth constraints suggest that 10 kilometers (33,000 feet) is the correct cruising altitude. It does not pay to go higher, because it is just as cold higher up. Conditions are optimal at the tropopause (the boundary between the troposphere and the stratosphere), where the highest density consistent with the quest for low outside temperatures is found. The cruising altitude is not left to the designer's discretion, but is determined by straightforward engineering logic.

A well-designed jetliner must fly a little slower than Mach 0.9 at a cruising altitude of 10 kilometers. For present purposes, let's choose a speed of 250 meters per second (900 kilometers per hour, 560 miles per hour, Mach 0.83). That's on the safe side of the absolute maximum. Now let's use equation 1, which shows how wing loading depends on density and airspeed:

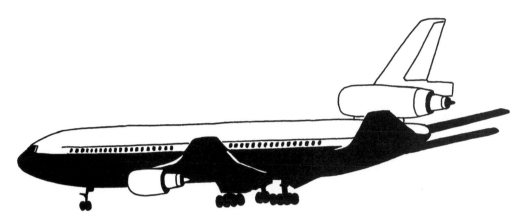

Douglas DC-10: $W = 2.56 \times 10^6$ N, $S = 368$ m^2, $b = 50$ m.

$W/S = 0.3\rho V^2$.

At an altitude of 10 kilometers, the air density ρ is 0.413 kilogram per cubic meter. (See table 6.) If we substitute $\rho = 0.413$ kg/m^3 and $V = 250$ m/s into equation 1, we find that the wing loading of our airplane, W/S, should be 7,740 N/m^2. What is the gross weight of a mundane, sensible airplane with that wing loading? Both oversize and undersize wings have their drawbacks, so let's stay in line with the main diagonal in figure 2:

$W/S = 47W^{1/3}$.

The result is $W = 4,470$ kilonewtons, or 447 tons. The wing area then becomes 578 square meters (about 6,200 square feet). It would have been easy to manipulate these numbers in such a way that the precise figures for the 747 would have been obtained. The 747-400 has a maximum takeoff weight of 394 tons and a wing area of 524 square meters (5,640 square feet). Our calculations would have come close if we had taken into account that the numerical coefficient in equation 1 should be somewhat smaller than 0.3 when computing the cruising speed of a jet, because of the improved engine efficiency at higher speeds. But that would have been nitpicking. Equations 1 and 3 are merely rules of thumb; great precision

Table 6 Atmospheric data: altitude h, temperature T, air density ρ, ratio between density at altitude and at sea level (ρ/ρ_0), ratio between cruising speed at altitude and at sea level (V/V_0), and speed of sound at altitude.

h (m)	T (°C)	ρ (kg/m³)	ρ/ρ_0	V/V_0	Speed of sound (m/sec)
0	15.0	1.225	1.000	1.00	340
1,000	8.5	1.112	0.908	1.05	333
2,000	2.0	1.007	0.822	1.10	333
3,000	−4.5	0.909	0.742	1.16	329
4,000	−11.0	0.819	0.669	1.22	325
5,000	−17.5	0.736	0.601	1.29	321
6,000	−24.0	0.660	0.539	1.36	317
7,000	−30.5	0.590	0.481	1.44	312
8,000	−37.0	0.525	0.429	1.53	308
9,000	−43.5	0.466	0.381	1.62	304
1,0000	−50.0	0.413	0.337	1.72	300
11,000	−56.5	0.364	0.297	1.83	295
12,000	−56.5	0.311	0.254	1.98	295
13,000	−56.5	0.266	0.217	2.15	295
14,000	−56.5	0.227	0.185	2.32	295

cannot be expected. Nor is great precision required to realize that the weight of an airplane designed to fly 900 kilometers per hour at an altitude of 10 kilometers should be approximately 400 tons.

The big jetliners of the 1960s, the Boeing 707 and the Douglas DC8, were *too small*. And they do seem small nowadays: the longest of the current 737s is almost as large as a 707, and carries even more passengers.

What about the Boeing 777, the Airbus A340, and the Airbus A380? The 777, the offspring of the 747, is on its way to maturity: the 777-300ER weighs 350 tons, and the next model may grow to 400 tons if the engineers at General Electric manage to improve their new engines a bit more. The two Airbuses are respectable competitors. As far as I am concerned, the size of all modern wide-body jets confirms my original analysis. Earlier jetliners were compromises between conflicting design criteria; the Boeing 747 was the first one that obeyed ruthless engineering logic. The fact that its successors and competitors are all of comparable size demonstrates unequivocally that the principles of flight have taken the decision out of human hands. Yet you shouldn't accept these conclusions unless I support them with further argumentation. The best way to argue the case is to see what happens if we deviate from the rules. After all, the main diagonal in the Great Flight Diagram allows plenty of scope for the ingenuity of aircraft designers. A designer can choose to move to the left or to the right of the main diagonal, giving a plane a lower or a higher wing loading than is typical for its weight.

Need a jetliner weigh 400 tons? Of course not; it is easy enough to design a smaller airplane. But a smaller airplane has a smaller wing loading if it has wings to suit its size, and a lower cruising speed to match. If you nonetheless insist on cruising at 560 miles per hour, you must fly abnormally high. The Boeing 737-300 weighs 57 tons; if it had ordinary wings for its size, its wing loading would be 3,900 newtons per square meter and its wing area 146 square meters (equation 3). At a design speed of 900 kilometers per hour, or 250 meters per second, the air density would have to be 0.2 kilogram per cubic meter (equation 1). The corresponding cruise altitude is 15 kilometers (49,000 feet). But a short-distance airliner cannot afford to climb to such an altitude. For this reason, the designers of the 737 had to compromise both on cruising speed and on altitude. There was an advantage to be gained with a higher-than-average wing loading (that is, with undersize wings). With a wing size of 105 square meters (see table 7), the wing loading of the 737 is in fact a little over 5,400 newtons per square meter—40 percent higher than average for airplanes of similar weight.

Why not produce a 737 with the same wing loading as a 747? The wings would be too small. The wings of a 57-ton plane with a wing loading equal to that of a 747 would be no larger than 77

Table 7 Dimensional data on popular airliners.

	Take-off weight, W (tons)	S (m²)	B (m)	Sea-level thrust, T (tons)	W/T	W/S	Seats
Airbus 380-800	560	845	80	4 × 36.3	3.86	6,627	550
Boeing 747-8	440	570	69	4 × 30.2	3.64	7,719	470
Boeing 747-400	395	524	65	4 × 25.7	3.84	7,538	421
Airbus 340-600	368	439	63	4 × 25.4	3.62	8,383	380
Boeing 747-200	352	511	60	4 × 21.3	4.13	6,888	387
Boeing 777-300ER	351	435	65	2 × 52.3	3.35	8,089	365
Boeing 777-200ER	297	428	61	2 × 42.5	3.49	6,939	305
Airbus 350-900	265	443	64	2 × 39.5	3.35	5,982	314
Boeing 787-9	245	370	60	2 × 31.1	3.94	6,622	250
Boeing 767-300	172	283	48	2 × 26.3	3.27	6,078	224
Boeing 707-320B	151	283	44	4 × 8.4	4.49	5,335	190
Boeing 737-900	85	125	35	2 × 12.4	3.43	6,800	180
Airbus 320	74	123	34	2 × 12	3.08	6,016	150
Boeing 737-300	57	105	29	2 × 9.1	3.13	5,429	124
Boeing 737-200	52	91	28	2 × 8.5	3.06	5,714	110
Fokker 100	43	93	28	2 × 6.7	3.21	4,624	107
Bombardier CL600	24	55	21	2 × 4.1	2.93	4,364	50
Embraer ERJ145	21	51	20	2 × 3.3	3.18	4,118	50

square meters (830 square feet), which is only half the median value for the plane's weight. A bird with undersize wings has a comparatively fat body, which creates additional air resistance and renders the design less economical than it could be. Moreover, a 737 with a wing loading equal to that of a 747 would need two-mile runways. It would not be able to land at smaller airports, where the standard runway is only a mile long. The 737 is intended for short distances, and a commuter plane unable to use regional airports makes no economic sense. Although the wing loading of a 737 is above average for its weight class, it is not excessive.

The Boeing 737 is a reasonable compromise between the desire to fly faster than is appropriate for its weight and the price to be paid for insisting on doing this. It is a sensible solution to conflicting design specifications, but it remains a compromise. Many versions of the 737 have been produced—some short, some with various stretched fuselages, some with winglets and some without. Because the whims of executives determine such matters, not engineering logic, airlines can place customized orders. Boeing willingly complies.

What about a much larger airplane—a 1,000-ton giant with a wing loading of 10,000 newtons per square meter? Yes, it would fit right on the trend line in figure 2. However, at that wing loading runways would have to be three miles long instead of two. Also, such a giant plane would have to fly lower if required to fly at 900 kilometers per hour. It would have to travel at an altitude where the air density is 0.53 kilograms per cubic meter (equation 1). But that corresponds to 8 kilometers (26,000 feet; see table 6), which is not high enough to be above the weather. Also, it is not cold enough at this height: only −37°C, not −56°C as in the stratosphere. The wing loading would have to be reduced in order to reach a cruising height of 10 kilometers and to maintain a decent landing speed. This means that the plane would need oversize wings, which would make it heavier than necessary and would reduce its payload capacity.

Boeing or Airbus?

The Airbus A380, with a takeoff weight of 560 tons, competes with the Boeing 747 for long-range traffic between major airports. Boeing has responded with a lengthened 747, called the 747-8. Both show signs of engineering compromises. Because airport restrictions limit the wingspans of jetliners to 80 meters, the wings of the A380 are not nearly as slender as those of a Boeing 777. Also,

Airbus A380: $W = 5,600$ kN, $S = 845$ m², $b = 80$ m.

neither the Boeing 747-8 nor the Airbus A380 is suited for the most powerful engine currently available, the General Electric GE90-115B.

The GE90-115B engine produces 115,000 pounds (512 kilonewtons, 52 tons) of thrust at takeoff. Almost all of the thrust is generated by an enormous fan up front, with a diameter of 3.25 meters (almost 11 feet). As in all modern fanjet engines, the engine itself (called the *core*) is much smaller in diameter than the fan. Most of the air sucked in by the fan bypasses the core: the "bypass ratio" is 10, meaning that 10 times as much air flows around the engine as into the compressor. The bypass ratio in the fanjet engines of the Boeing 707 was as small as 1:1. In the engines of the Boeing 747 it increased to 4:1. The recent step forward to a bypass ratio of 10:1 reduces fuel consumption but tightens the operational envelope. Don't fly these engines too fast or too slow! Present-day fanjet engines can be thought of as propjets that hide their propeller in a casing. The casing makes the propeller more efficient and substantially reduces unwanted engine noise. At takeoff, the fan swallows 3,600 pounds of air per second. The air pressure at the exit of the compressor, which is the entrance to the combustion chambers, is 40 times the ambient air pressure. Compare this compression ratio to that of a gasoline engines (typically only 8:1) or that of a diesel engines (about 20:1). When the exhaust gases hit the first turbine stage, the temperature has risen to almost 1,400°C (that's 2,500°F). Such extreme temperatures require extremely hardy turbine

Fokker F-100: $W = 4.3 \times 10^5$ N, $S = 94$ m^2, $b = 28$ m.

blades, but the efforts and expenses involved (a single engine of this size costs about $20 million) are paid back in improved engine efficiency, lower specific fuel consumption, and longer range. Eighteen-hour nonstop flights have become possible.

The main reason for the popularity of high-bypass-ratio jet engines is their low fuel consumption. At cruising altitude, each GE90-115B engine consumes about 5,000 liters per hour, and produces 10 tons of thrust at an airspeed of 900 kilometers per hour. Force times speed equals power, so in metric units we have 100 kilonewtons $\times$ 250 meters per second, or 25 megawatts (33,000 horsepower). The fuel consumption then computes at 0.15 liter per horsepower per hour. Your car, which needs about 20 horsepower at a speed of 100 kilometers per hour and uses about 7 liters per hour at that speed, consumes about 0.3 liter per horsepower per hour. This is in line with other numbers we have encountered. The thermal efficiency of the piston engine in your car is only 25 percent, but a high-bypass-ratio jet engine, assisted by very low outside air temperatures and high turbine temperatures, manages

50 percent! Both for the jet and for the car, the heat of combustion computes as 36 megajoules per liter (about 45 megajoules per kilogram), corresponding fairly well with the data in table 3.

What are the design parameters of airplanes powered by the GE90-115B? Designing a four-engine jetliner around these engines is straightforward. Four times 52 tons of thrust equals 208 tons. The takeoff weight fitting this number is 3.6 × 208 tons, that is 750 tons. (For the weight-to-thrust ratio of large airliners, see table 7.) We limit the wing loading to 7,500 newtons per square meter, so the wing area needed is 1,000 square meters. We want a slender wing to minimize induced drag, so we should choose an aspect ratio of 9 (like that of a Boeing 777). The wingspan then becomes 95 meters. That would be one beautiful airplane, good for 700 passengers, but unfit for docking at present airport gates.

Now let's take this argument one step further. Suppose we want to build a modern competitor for the Antonov AN225, with six engines of 52 tons thrust each. The takeoff weight of such a plane would be 1,120 tons. Since the wing loading has to be kept at 7,500 newtons per square meter, a wing of 1,500 square meters, with a span of 116 meters, would be needed. This plane could carry 1,000 passengers, but bigger is not necessarily better. For the time being, I don't see any aircraft company daring to build a 1,000-ton, 1,000-passenger airliner. For as long as I can remember, designers have floated speculations about "super-jumbos," but the largest airliner now flying, the bloated Airbus A380, is at best a "mini-super."

The logical alternative is a twin-engine jetliner with each engine delivering 52 tons of thrust and with a takeoff weight of 374 tons. Keeping the wing loading at 7,500 newtons per square meter, we obtain a wing area of 500 square meters. Something special is happening here. Our off-hand calculation produces numbers that compare well with those for the 777-300ER, but also with those of the 747-200 and the Airbus 340. (See table 7.) This is what biologists call "convergence." All the long-range airliners that have survived in the evolutionary struggle for success have a weight and a wing

Boeing 737-700: $W = 600$ kN, $S = 125$ m^2, $b = 34$ m.

area not unlike those of the original Boeing 747. Still, if I had to choose among the 747, the Airbus A380, the Airbus A340, and the 777, I would pick the 777, because it has two engines rather than four.

In Praise of the 747

The glory of the Boeing 747 is fading. The world's major airlines are slowly phasing it out. About 1,400 of the planes are still around; all of them will gradually be shifted into charter and freight service. Not many stretched 747s will be built. But all descendants and competitors of the 747 are testimony to its incredible success. No wonder some designers dream of a "747-twin" that would retain the old airframe but would replace the four original engines with two of the current breed. However, I am not sure that is a good idea.

The 747 remains one of the great engineering wonders of the world, like the pyramids of Egypt, the Eiffel Tower, or the Panama Canal. No longer the largest wide-body airliner in the world, but

Boeing 737-300: $W = 5.7 \times 10^5$ N, $S = 105$ m^2, $b = 29$ m.

surely the most successful one, it incorporates everything one can reasonably demand of a mode of transportation: reliability, ease of maintenance, productivity, fuel economy, speed, and ample cargo space. For its weight, the 747's wing loading is quite ordinary. Nor need the size of the 747 impress us. From the viewpoint of economy of scale, it is a happy coincidence that the optimal solution necessitates a large and heavy airplane; however, the size of the 747 is not a matter of choice, even though it may have seemed that way to Mr. Trippe and the Boeing design team.

A single 747 flying back and forth between Amsterdam and New York produces at least 2 million seat-miles a day, based on 300+ seats and a one-way distance of 3,600 statute miles. Even with maintenance and periodic inspections, a 747 makes more than 300 round trips a year, so its annual production, not including freight, is 600 million seat-miles. Ten 747s match the entire traffic volume of the Netherlands State Railways. Admittedly the Netherlands is a small country, but nevertheless several hundred coaches and commuter trains are required to achieve that performance. In the United States, an Amtrak train making the two-day run between Chicago and San Francisco with 400 passengers aboard accounts, in theory, for less than half a million seat-miles a day, and in practice delays, maintenance, and time-consuming turnaround procedures cut that in half: the net rate is only 200,000 seat-miles a day. The French high-speed train, the TGV, fares somewhat better. The 250-mile run between Paris and Lyon takes 2 hours and carries 500 people. Making three round trips a day, a TGV, with almost twice as many seats as a 747, produces more than 700,000 seat-miles a day—but that is only one-third of a 747's productivity.

A new 747 costs roughly $200 million. The first owner writes this cost off over 10 years. After 10 years, though, a 747 still has plenty of life in it. Since the major airlines are now phasing out their 747s, you may want to profit from the current buyers' market. Let's assume you can buy a ten-year old 747 for $100 million. Average depreciation and interest over the first 10 years are estimated at $10 million and $5 million per year, respectively, for a total of $15 million per year. One-third of that amount must be earned by carrying freight, leaving roughly $10 million to be recovered by selling 600 million seat-miles. This works out at 2 cents per seat-mile. Compare that with your car. The Consumers Union estimates automobile interest and depreciation at about 30 cents per mile. With two people in a car on average, the cost is thus 15 cents per passenger-mile—considerably more expensive than an airplane. Trains are not very economical on this score, either. Per seat-mile, the direct energy consumption of a train is half that of a car (see chapter 2), but the other costs are disappointing. Railroads must maintain an extensive infrastructure and must invest heavily in their inefficiently utilized rolling stock. In order to break even in their passenger operations, railroads the world over depend on massive state subsidies. As a rule of thumb, the price of a railroad ticket covers only half of the real cost.

It is easy to see why the great ocean liners were doomed as soon as jetliners appeared on the transatlantic market. A ship of some stature easily costs $500 million. Let's assume that a round trip between Southampton and New York takes 2 weeks; that allows 20 round trips a year if we exclude the winter season. With 1,500 passengers on board, an ocean liner produces 200 million passenger-miles a year. If we estimate interest and depreciation optimistically at $50 million a year, this comes to 25 cents per passenger-mile—10 times the capital expenses of a 747. The fuel consumption of an ocean liner isn't very favorable either: an ocean liner burns 90 gallons of fuel oil per mile, so the carbon footprint of each passenger is worse than that of each person in a jetliner. Low-speed travel is relaxing, but it remains unprofitable when you look at total productivity. A slow mode of transportation makes it difficult to recoup one's investment. This is what drove ocean liners into the luxury cruise business, this is why airships (zeppelins) will survive only in tiny niche markets.

All wide-body jets are unequaled as freight carriers. Let's compare freight against passengers. With luggage and meals included, passengers account for an average of 100 kilograms each. On intercontinental flights economy-class passengers pay approximately 10 cents per mile. Of this, 2 cents is for fuel and 2.5 cents for capital expenses. Converted to weight, the passenger tariff becomes roughly a dollar per ton-mile. But freight doesn't require flight attendants, chairs, pillows, blankets, toilets, kitchens, and meals. It should therefore be possible to offer freight service at approximately half the price per ton-mile. Indeed, the going rate for intercontinental airfreight is about 50 cents per ton-mile, increasing to a dollar per ton-mile for fresh-cut flowers, vegetables, and other perishables, and with additional surcharges for horses, elephants, and day-old chicks.

Imagine you are a fashion buyer at Macy's in New York or Harrods in London, and you need an extra supply of some suddenly hot-selling jeans from a supplier in Asia. Now, 2,000 pairs of denim trousers, at roughly a pound each, weigh about a ton. Any Boeing or Airbus can carry that order 10,000 miles at a cost of $6,000, or $3 per pair. Your department store can recover this expense with ease by increasing the price of the jeans from $49 to $59.

Gladioli from South Africa can't lie flat; they have to be transported upside down to prevent their tips from becoming crooked. Roses from Argentina find their way to American flower shops on the same day. Freesias from Israel are auctioned off at dawn near Amsterdam and are on their way to Chicago before noon. The sky is literally the limit in the variety of products that are flown around the world. Take off-season string beans: 1,200 miles between Tunisia and Holland, at a rate of 50 cents per ton-mile, make the transportation cost only 30 cents per pound. In the middle of winter I certainly would not mind paying that extra charge for a special treat.

Though airfreight has a reputation of being expensive, the differences are not as great as one might think. FedEx and UPS offer surface rates not much below the rates for regular airfreight. Indeed, FedEx owns a large fleet of cargo planes, and it wouldn't if that weren't profitable. Substantially lower rates are offered by long-distance trucking companies. Bulk shipments of ore and grain by rail are cheaper still, and barges on the Great Lakes charge only 2 cents per ton-mile or even less. For the time being, therefore, one should not expect to find planes being loaded with ore or gravel. Nevertheless, the potential of airfreight should not be underestimated. Once fully depreciated 747s are bought up by freight and charter operators, the rate may drop to 30 cents per ton-mile. The first 747 entered service with Pan American in 1969. There are plenty superannuated 747s are around today. Just look around the tarmac the next time you are waiting for a connecting flight at Atlanta or Pittsburgh: the far ends of the apron are cluttered with unmarked airliners painted in unattractive colors, with their windows riveted shut.

The Boeing 747 was designed with the North Atlantic in mind. This was the corridor with the most business, the fiercest competition, and the most potential income 40 years ago, and it is still the densest corridor today. By a stroke of luck, it also happened to be the run that provides the most efficient flight schedules. In the 1950s, a propeller plane took an average of 14 hours to make the crossing; a day later it began its return trip. Nowadays a westbound flight takes about 8 hours, and 3 hours later the plane is on its way back, to land in London, Amsterdam, or Frankfurt 7 hours later. Just 18 hours after its departure, a wide-body jet is back home, in time for the early cleaning and maintenance shifts. An excellent routine: no clumsy personnel from outside contractors, no overtime

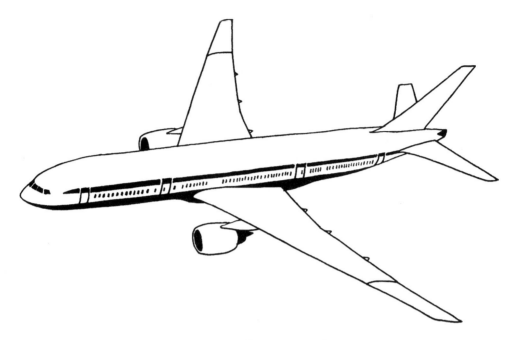

Boeing 777-200: $W = 2.4 \times 10^7$ N, $S = 428$ m^2, $b = 61$ m.

payments, and a fixed pattern of home port maintenance running like clockwork. Such a routine helps to keep costs down.

On the longer runs between Europe and the United States there is little time to spare. Amsterdam–San Francisco must be completed within 10 hours, or else a fixed schedule with the plane back home every morning becomes impossible. That journey is 5,000 miles, 1,500 miles longer than the Amsterdam–New York run. A travel time of 10 hours demands a cruising speed of at least 500 miles per hour to allow time for taxiing, taking off, waiting in the holding pattern, and landing.

Airline traffic across the Pacific has become a lot denser in recent years, and will continue to grow rapidly as the Chinese economy expands. This corridor is served best by ultra-long-range airliners capable of flying at least 8,000 miles nonstop. All current wide-body jets can fill this bill, but not without some drawbacks. For one, two cockpit crews are needed, because pilots who have been

Airbus A380: $W = 5600$ kN, $S = 845$ m^2, $b = 80$ m.

flying 14 hours or more are likely to be too sleepy to make a safe landing. Also, return flights on the same day are out of the question, which makes scheduling and utilization more difficult.

Lose Weight, Gain Altitude

A jetliner's weight decreases as it consumes its fuel, but the wings cannot change size. Therefore, the wing loading decreases. If the plane remained at the same altitude, so that the air density remained constant, the decreasing wing loading would require the plane to slow down, thus affecting the schedule. Moreover, the efficiency of the jet engines would suffer, for they do not perform optimally at lower speeds. As its weight decreases, a plane must find more rarefied air in order to maintain its speed. If its wings have become too large for flying at an altitude of 10 kilometers, it moves up a kilometer. The engines don't mind: when the weight decreases, so does the drag if the finesse remains the same. Less drag means that the engines need to deliver less thrust. It does not matter, therefore, that as the cruising altitude increases the engines breathe somewhat thinner air.

On a long intercontinental flight, say from Tokyo to Amsterdam, a Boeing 747-400 might start out cruising at an altitude of 9,000 meters (30,000 feet) and a weight of 380 tons. Thirteen hours later, when it begins its descent toward Amsterdam over Berlin, its weight has decreased by one-third, to 250 tons. In the meantime, it has climbed to an altitude where the air density is also 66 percent of its initial value: 12,100 meters (40,000 feet). Each hour it must climb roughly 800 feet in order to compensate for the fuel burned.

Epilogue

More than 2,000 years ago the Romans invested heavily in a great highway system. The empire needed a good road infrastructure in order to maintain unity and suppress local conflicts. The Roman highways were engineered so well that their remnants can still be found all over Europe. In seventeenth-century Holland, canals were dug to promote passenger traffic on a regular timetable, independent of fog and wind. A Haarlem banker could commute to Amsterdam, do his business, and be back home in time for dinner. The mobility of the merchant class helped to keep the United Provinces together. In the eighteenth century, England and France built extensive canal systems, judging a massive investment in infrastructure to be the best preparation for the future. As often happens, this investment came too late: by the time the canal system became fully operational a century later, railroads made the expensive canal system obsolete. The first transcontinental railroad in the United States dates from 1869, when the Union Pacific met the Central Pacific at Promontory Point in Utah. Nearly a century later, this infrastructure was supplemented by the interstate highway system, originally conceived, like its Roman predecessor, for rapid movement of military materiel. Over and over again, statesmen have learned that adequate transportation facilities are necessary to keep a country united.

When Marshall McLuhan dreamed of the "global village," he was thinking primarily of telecommunications. He felt that the incessant chatter on long-distance party lines would inevitably bring people closer together. We can't blame him for not forecasting the cell-phone and Internet revolutions that have changed the way people all over the world communicate. But McLuhan also did not pay sufficient attention to the continuing need for physical mobility. Ultimately, cell phones and e-mail are not enough. The time comes when you want to see the Grand Canyon or the Pyramids for yourself, meet and talk to people from other continents, or sip beer in the backyard of your international business partner. Jetliners have become the indispensable commuter buses of the global village.

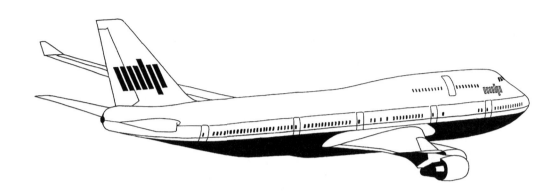

Appendix   Flight Data for Migrating Birds

Birds are listed in order of ascending weight. The weight W is given in newtons, the wing area S in square meters, and the wingspan b in meters. The cruising speed (meters per second) calculated with $W/S = 0.38 V^2$ is listed as V_c, and the observed migration speed in level flapping flight as V_m. All data are taken from the following papers: B. Bruderer and A. Boldt, Flight characteristics of birds: I. Radar measurements of speeds, *Ibis* 143(2001): 178–204; T. Alerstam et al., Flight speeds among bird species: Allometric and phylogenetic effects, *PLoS Biology* 5(2007): 1656–1662.

In general, small birds not on migration fly substantially slower than given here, at speeds comparable to those calculated. Very large birds, such as swans, fly somewhat slower than calculated, at speeds which require the least muscle power. Migrating waders, such as plovers, knots, and godwits, fly faster than most geese. Ducks are fast, too. Gulls, terns, and eagles, on the other hand, seem to take it easy, even on migration.

Several species listed here prefer to soar if they have a chance. Storks and eagles do not cross the Mediterranean Sea unless pressed for time or seduced by strong tailwinds, because there are no thermals over open sea. They travel via Gibraltar or Israel. The net cross-country speed of soaring birds is relatively low, if only because thermals occur only in daytime.

	W	S	b	V_c	V_m
Coal tit *Parus ater*	0.09	0.0073	0.18	5.7	10.6
Siskin *Carduelis spinus*	0.14	0.0076	0.21	7	10.6
Sand martin *Riparia riparia*	0.15	0.0096	0.27	6.4	14.3
House martin *Delichon urbica*	0.15	0.0104	0.29	6.2	9.7
Barn swallow *Hirundo rustica*	0.16	0.0136	0.32	5.6	10
Yellow wagtail *Motacilla flava*	0.18	0.0103	0.26	6.8	12.7
Great tit *Parus major*	0.19	0.0109	0.23	6.8	13.6
White wagtail *Motacilla alba*	0.21	0.0126	0.26	6.8	14.1
Tree pipit *Anthus trivialis*	0.22	0.0126	0.27	6.8	12.7
Swift *Apus apus*	0.38	0.0168	0.4	7.7	9.7
Skylark *Alauda arvensis*	0.39	0.0207	0.35	7.0	15.1
Dunlin *Calidris alpina*	0.54	0.0156	0.36	9.5	15.3
Redwing *Turdus iliacus*	0.61	0.0233	0.36	8.3	13.8
Ringed plover *Charadius hiaticula*	0.64	0.0179	0.41	9.7	19.5
Song thrush *Turdus philomenos*	0.68	0.0218	0.36	9.1	11
Alpine swift *Apus melba*	0.78	0.0304	0.57	8.2	12.6
Starling *Sturnus vulgaris*	0.83	0.0244	0.38	9.5	16.2

	W	S	b	V_c	V_m
Arctic tern *Sterna paradisaea*	1.1	0.0571	0.8	7.1	10.9
Turnstone *Arenaria interpres*	1.11	0.0252	0.47	10.8	14.9
Mistle thrush *Turdus viscovorus*	1.14	0.0333	0.44	9.5	11.9
Red knot *Calidris canutus*	1.28	0.0286	0.5	10.9	20.1
Eurasian jay *Garrulus glandarius*	1.62	0.0644	0.54	8.1	12.9
Greenshank *Tringa nebularia*	1.74	0.0406	0.61	10.6	12.3
Kestrel *Falco tinninculus*	2.03	0.0708	0.73	8.7	10.1
Grey plover *Pluvialis squatarola*	2.19	0.042	0.62	11.7	17.9
Northern lapwing *Vanellus vanellus*	2.19	0.0744	0.75	11.7	17.9
Hobby *Falco subbuteo*	2.38	0.0667	0.73	9.7	11.3
Jackdaw *Corvus monedula*	2.45	0.0684	0.65	9.7	12.5
Sparrow hawk *Accipiter nisus*	2.77	0.0768	0.67	9.7	11.3
Black-headed gull *Larus ridibundus*	2.83	0.0976	0.97	8.7	11.9
Long-tailed skua *Stercorarius longicaudus*	2.97	0.0891	1.01	9.4	13.6
Bar-tailed godwit *Limosa lapponica*	3.18	0.052	0.73	12.7	18.3
Common teal *Anas crecca*	3.48	0.0428	0.59	14.6	19.7
Eleonora's falcon *Falco eleonora*	3.87	0.104	0.95	9.9	12.8

	W	S	b	V_c	V_m
Kittiwake *Rissa tridactyla*	4.08	0.0953	0.96	10.6	13.1
Common gull *Larus canus*	4.11	0.1246	1.11	9.3	13.4
Hen harrier *Circus cyaneus*	4.33	0.157	1.1	8.5	9.1
Arctic skua *Stercorarius parasiticus*	4.38	0.118	1.06	9.9	13.8
Rook *Corvus frugilegus*	4.88	0.138	0.91	10.4	13.5
Wood pigeon *Columba palumbus*	4.91	0.0824	0.75	12.4	16.3
Carrion crow *Corvus corone*	5.66	0.138	0.91	10.4	13.5
Marsh harrier *Circus aeruginosus*	6.53	0.204	1.16	9.2	11.2
Lesser black-backed gull *Larus fuscus*	7.19	0.193	1.34	9.7	13.1
Honey buzzard *Pernis apivorus*	7.78	0.247	1.26	9.1	12.5
Peregrine falcon *Falco peregrinus*	7.89	0.1257	1.02	12.9	12.1
Buzzard *Bureo buteo*	8.85	0.269	1.24	9.3	11.6
Rough-legged buzzard *Buteo lagopus*	9.43	0.332	1.35	8.6	10.5
Red kite *Milvus milvus*	10.12	0.325	1.66	9.1	12
Pintail *Anas acuta*	10.24	0.0879	0.9	17.5	20.6
Mallard *Anas platyrhynchos*	10.82	0.106	0.88	16.4	18.5
Herring gull *Larus argentatus*	11.42	0.197	1.34	12.4	12.6

	W	S	b	V_c	V_m
Raven *Corvus corax*	11.5	0.247	1.21	11.1	14.3
Brent goose *Branta bernicla*	13.1	0.113	1.01	17.4	17.7
Lesser spotted eagle *Aquila pomarina*	13.9	0.515	1.47	8.4	11.7
Grey heron *Ardea cinerea*	14.4	0.372	1.73	10.1	12.5
Goosander *Mergus merganser*	14.9	0.077	0.93	22.6	19.7
Red throated diver *Gavia stellata*	15.1	0.089	1.04	21.1	18.6
Osprey *Pandion haliaetus*	15.8	0.32	1.6	11.4	13.3
King eider *Somateria spectabilis*	15.9	0.108	0.93	19.7	16
Great black-backed gull *Larus marinus*	16.7	0.288	1.67	12.4	13.7
Barnacle goose *Branta leucopsis*	17.1	0.115	1.08	19.8	17
Common eider *Somateria mollissima*	20.2	0.131	0.98	20.1	17.9
Cormorant *Phalacrocorax carbo*	22.3	0.224	1.4	16.1	15.2
Black-throated diver *Gavia arctica*	25.4	0.12	1.2	23.6	19.3
White-fronted goose *Anser albifrons*	25.8	0.184	1.41	19.2	16.1
Bean goose *Anser fabilis*	30.4	0.268	1.62	17.3	17.3
Greylag goose *Anser anser*	33.3	0.308	1.55	16.9	17.1
White stork *Ciconia ciconia*	34.3	0.533	1.91	13	16

	W	S	b	V_c	V_m
Canada goose *Branta canadensis*	36.3	0.372	1.69	16	16.7
Golden eagle *Aquila chrysaetos*	40.7	0.597	2.03	13.4	11.9
Common crane *Grus grus*	56.1	0.586	2.22	15.9	15
Tundra swan *Cygnus columbianus*	66.4	0.461	1.98	19.5	18.5
Whooper swan *Cygnus cygnus*	86.9	0.605	1.98	19.5	18.5
Mute swan *Cygnus olor*	106	0.65	2.3	20.7	16.2

Bibliography

T. Alerstam, *Bird Migration*. Cambridge University Press, 1990.

R. M. Alexander, *Exploring Biomechanics*. Freeman, 1992.

R. M. Alexander, *Principles of Animal Locomotion*. Princeton University Press, 2003.

D. F. Anderson and S. Eberhardt, *Understanding Flight*. McGraw-Hill, 2000.

J. D. Anderson, *Introduction to Flight*. McGraw-Hill, 1989.

J. D. Anderson, *A History of Aerodynamics*. Cambridge University Press, 1997.

A. Azuma, *The Biokinetics of Flying and Swimming*. Springer, 1992.

R. Bach, *Jonathan Livingston Seagull*. Avon, 1970.

G. I. Barenblatt, *Dimensional Analysis*. Gordon and Breach, 1987.

A. K. Brodsky, *The Evolution of Insect Flight*. Oxford University Press, 1994.

R. Burton, *Bird Migration*. Aurum, 1992.

J. J. Corn, *The Winged Gospel: America's Romance with Aviation*. Oxford University Press, 1983.

T. D. Crouch, *Wings: A History of Aviation*. Norton, 2003.

P. J. Currie, *The Flying Dinosaurs*. Red Deer College Press, 1991.

S. Dalton, *The Miracle of Flight*. McGraw-Hill, 1977.

R. Dudley, *The Biomechanics of Insect Flight*. Princeton University Press, 2000.

N. Elkins, *Weather and Bird Behavior*. Poyser, 1983.

J. Elphick, *The Atlas of Bird Migration*. Marshall, 1995.

M. French, *Invention and Evolution*, Cambridge University Press, 1994.

R. G. Grant, *Flight: 100 Years of Aviation*. Duxford Kindersley, 2002.

M. Grosser, *Gossamer Odyssey*. Zenith, 2004.

B. Gunston, *The Development of Jet and Turbine Aero Engines*, Zenith, 2006.

B. Gunston, *World Encyclopedia of Aero Engines*. Sutton, 2006.

R. P. Hallion, *Taking Flight*. Oxford University Press, 2003.

T. A. Heppenheimer, *A Brief History of Flight*. Wiley, 2001.

K. Huenecke, *Jet Engines*. Airlife, 1997.

C. Irving, *Wide-Body: The Making of the 747*. Hodder & Stoughton, 1993.

P. Kerlinger, *How Birds Migrate*. Stackpole, 1995.

P. Kerlinger, *Flight Strategies of Migrating Hawks.* University of Chicago Press, 1989.

J. Lienhard, *The Engines of Our Ingenuity.* Oxford University Press, 2000.

R. Lorenz, *Spinning Flight: Frisbees, Boomerangs, More.* Copernicus, 2006.

U. M. Norberg, *Vertebrate Flight.* Springer, 1990.

G. Norris & M. Wagner, *Airbus A380.* Zenith, 2005.

G. S. Paul, *Dinosaurs of the Air.* Johns Hopkins University Press, 2002.

C. J. Pennycuick, *Animal Flight.* Edward Arnold, 1972.

C. J. Pennycuick, *Bird Flight Performance.* Oxford University Press, 1989.

C. J. Pennycuick, *Newton Rules Biology.* Oxford University Press, 1992.

C. J. Pennycuick, *Modelling the Flying Bird.* Academic, 2008.

H. Petroski, *To Engineer Is Human.* Vintage, 1992.

H. Petroski, *Invention by Design.* Harvard University Press 1996.

H. Petroski, *Success through Failure—The Paradox of Design.* Princeton University Press, 2006.

G. Rueppell, *Bird Flight.* Van Nostrand Reinhold, 1977.

K. Sabbach, *777: Twenty-First Century Jet.* Scribner, 1996.

K. Schmidt-Nielsen, *Scaling: Why Is Animal Size So Important?* Cambridge University Press, 1997.

E. J. Slijper and J. M. Burgers, *The Art of Flying in the Animal Kingdom.* In Dutch. Brill, 1950.

Blanche Stillson, *Wings: Insects, Birds, Man.* Gollancz, 1955.

N. Shute, *No Highway.* Morrow, 1948.

J. Sutter, *747: Creating the World's First Jumbo Jet.* Collins, 2006.

E. S. Taylor, *Dimensional Analysis for Engineers.* Oxford University Press, 1974.

M. J. H. Taylor, *Jane's Encyclopedia of Aviation.* Crescent, 1995.

J. D. Terres, *How Birds Fly.* Stackpole, 1994.

J. J. Videler, *Avian Flight.* Oxford University Press, 2005.

W. G. Vincenti, *What Engineers Know and How They Know It.* Johns Hopkins University Press, 1993.

S. Vogel, *Life in Moving Fluids.* Princeton University Press, 1994.

S. Vogel, *Life's Devices.* Princeton University Press, 1988

S. Vogel, *Comparative Biomechanics.* Princeton University Press, 2003.

P. P. Wegener, *What Makes Airplanes Fly?* Springer, 1997.

R. Whitford, *Evolution of the Airliner.* Crowood, 2007.

D. G. Wilson, *Bicycling Science.* MIT Press, 2004.

Index